Arctic Geopolitics, Media and Power

T0187900

Arctic Geopolitics, Media and Power provides a fresh way of looking at the potential and limitations of regional international governance in the Arctic region.

Far-reaching impacts of climate change, its wealth of resources and potential for new commercial activities have placed the Arctic region into the political limelight. In an era of rapid environmental change, the Arctic provides a complex and challenging case of geopolitical interplay. Based on analyses of how actors from within and outside the Arctic region assert their interests and how such discourses travel in the media, this book scrutinizes the social and material contexts within which new imaginaries, spatial constructs and scalar preferences emerge. It places ground-breaking attention to shifting media landscapes as a critical component of the social, environmental and technological change. It also reflects on the fundamental dilemmas inherent in democratic decision making at a time when an urgent need for addressing climate change is challenged by conflicting interests and growing geopolitical tensions.

This book will be of great interest to geography academics, media and communication studies and students focusing on policy, climate change and geopolitics, as well as policy-makers and NGOs working within the environmental sector or with the Arctic region.

Annika E Nilsson is a researcher at KTH Royal Institute of Technology. Her work focuses on the politics of Arctic change and communication at the science–policy interface. Nilsson was previously at the Stockholm Environment Institute.

Miyase Christensen is Professor of Media and Communication Studies at Stockholm University and is an affiliated researcher at KTH the Royal Institute of Technology. Christensen's research focuses on environmental communication; technology-social change; and politics of mediation.

Routledge Geopolitics Series
Series Editors
Klaus Dodds, Professor of Geopolitics at the Department of Geography, Royal Holloway University of London, Egham Surrey UK. k.dodds@rhul.ac.uk

Reece Jones Professor of Geography at the Department of Geography, University of Hawai'i at Manoa, Hawai'i, USA. reecej@hawaii.edu

Geopolitics is a thriving area of intellectual enquiry. The *Routledge Geopolitics Series* invites scholars to publish their original and innovative research in geopolitics and related fields. We invite proposals that are theoretically informed and empirically rich without prescribing research designs, methods and/or theories. Geopolitics is a diverse field making its presence felt throughout the arts and humanities, social sciences and physical and environmental sciences. Formal, practical and popular geopolitical studies are welcome as are research in areas informed by borders and bordering, elemental geopolitics, feminism, identity, law, race, resources, territory and terrain, materiality and objects. The series is also global in geographical scope and interested in proposals that focus on past, present and future geopolitical imaginations, practices and representations.

As the series is aimed at upper-level undergraduates, graduate students and faculty, we welcome edited book proposals as well as monographs and textbooks which speak to geopolitics and its relationship to wider human geography, politics and international relations, anthropology, sociology, and the interdisciplinary fields of social sciences, arts and humanities.

Popular Geopolitics
Plotting an Evolving Interdiscipline
Edited by Robert A. Saunders and Vlad Strukov

Arctic Geopolitics, Media and Power
Annika E Nilsson and Miyase Christensen

For more information about this series, please visit: www.routledge.com/Routledge-Geopolitics-Series/book-series/RFGS

Arctic Geopolitics, Media and Power

**Annika E Nilsson and
Miyase Christensen**

Routledge
Taylor & Francis Group
LONDON AND NEW YORK

First published 2019
by Routledge
2 Park Square, Milton Park, Abingdon, Oxon OX14 4RN

and by Routledge
605 Third Avenue, New York, NY 10017

First issued in paperback 2020

Routledge is an imprint of the Taylor & Francis Group, an informa business

British Library Cataloguing-in-Publication Data
A catalogue record for this book is available from the British Library

Library of Congress Cataloging-in-Publication Data
A catalog record has been requested for this book

ISBN 13: 978-0-367-72836-6 (pbk)
ISBN 13: 978-0-367-18982-2 (hbk)

Typeset in Times New Roman
by Swales & Willis Ltd, Exeter, Devon, UK

To our children – Dane and Lara – and their generation

Contents

Figures

Tables

Preface and acknowledgements

This book and the research project with which it is connected have their origin in a collaboration led by the Division of History of Science, Technology and Environment at KTH Royal Institute of Technology, which resulted in the book *Media and the Politics of Arctic Climate Change: When the Ice Breaks* (Palgrave MacMillan, 2013). Its interdisciplinary context created momentum for thinking further about how the Arctic is mediated through a combination of science, politics and journalistic accounts, and how this mediation in turn affects visions of and decision making about the future. As we write this preface in December 2018, that future is unfolding, most notably in the accelerated warming of the Arctic and the recurring news about record high temperatures and lack of snow and ice, paralleled by reports that global emissions of greenhouse gases are rising rather than declining in line with the political promise made in the 2015 Paris Agreement. Meanwhile, the Arctic is evermore the focus of explorations for oil and gas, among other things. Like the rest of the world, the region has yet to engage seriously with the large-scale energy transformation that is necessary not only for the Arctic but also to safeguard human well-being globally.

The only way to change the current deadlock is to make decisions, at all levels in society, that a different future than that of rapid global warming is more desirable, and to act with this in mind. Regions such as the Arctic could potentially take a lead in initiatives and actions. The geopolitical interests of powerful states play a role in whether such a vision is realistic, but are not the only factors to consider. The future can also be shaped by new voices that are making themselves heard thanks to a shifting media landscape. Our interest in this interplay between decision making, visions and narratives brought us together to write this book.

Inspiration has come from the many colleagues and friends with whom we have worked over the years. In addition to colleagues at KTH, this includes our respective home institutions while working on this project—the Stockholm Environment Institute and the Department of Media Studies at Stockholm University—and the wider context of projects in which we have been engaged in the Arctic and elsewhere. We are especially grateful for the inspiration and insights shared by informants and interviewees and by the participants in two events: the workshop 'Global Arctic–Regional Governance: Visions and Mediations of Northern Spaces' held in Stockholm on 6–7 December 2017, and the panel discussion

'Arctic Journalism at the Crossroad of Technology, Economy and Politics' held at the 2018 Arctic Circle Assembly in Reykjavik. We also wish to thank Tom Buurman and Ekaterina Klimenko for their support with the empirical analyses of media coverage, and Andrew Mash for his support with editing.

The work has been made possible by funding from the Swedish Research Council Formas under the project *Arctic governance and the questions of 'fit' in an era of globally transformative change: A critical geopolitics of regional international cooperation* (contract 2011-2014-1020).

Finally, we want to thank our spouses for always being there, and our children for making it necessary and urgent to think about how we work today to shape their future.

Annika E Nilsson and Miyase Christensen
Stockholm, December 2018

1   The regional? Mediation, scale and power

We see Earth from space in a projection that highlights the northern polar region and where water rather than sea ice dominates. At times, potential or real shipping lanes are superimposed onto the image. At other times, the image depicts the latest claims to mineral rights under the continental shelf that might become accessible once the ice recedes even further and new offshore technologies have been developed. Other images focus on military capacity, in a region with a three-decade record of peaceful international cooperation but an even longer history as a heavily militarized meeting point between east and west.

At first shocking in their stark new reality, such images have quickly become the new 'normal'. These images appear in the major news media outlets, as part of expert reports, and in scientific presentations. The future of the Arctic region (see Figure 1.1) is also discussed in a plethora of forums where interested parties from around the world gather to claim their right to co-shape this vast space. Meanwhile, the approximately four million people living in the region are starting to make their own voices heard. Such interventions come from indigenous peoples who assert their rights to land, resources and knowledge, as well as their cultural identities in all their complexities. This is visible in explicit messaging through the media, particularly regional outlets, and at political forums, such as the Conference of the Parties to the United Nations Framework Convention on Climate Change, and through popular communication channels, such as film and music. The award-winning movie *Sami Blood* is one such example. The many expressions of Saami *joik* and *joik*-inspired music also reach a broad range of listeners in the mainstream media. Meanwhile, local elected politicians are creating new links across national borders, asserting their special role in the delivery of essential public services and ensuring that their communities are resilient and sustainable in the long term (Declaration of Arctic Mayors, 2017; Kristoffersen, 2017).

Over the past decade, the images and narratives that circulate through traditional news outlets have been accompanied by web-based specialist news services providing daily updates about the circumpolar north that reach readers both within and far beyond the Arctic. As fibre optic connections improve, social media are becoming more prominent in spreading individual local stories across the region and to an audience that might have had no previous connection

— Arctic bioclimate zone, north of treeline

━ Arctic Monitoring and Assessment Programme (AMAP)

--- Conservation of Arctic Flora and Fauna (CAFF)

····· Arctic Human Development Report (2004)

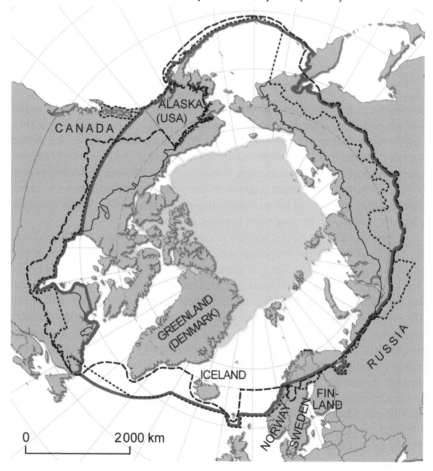

Figure 1.1 Map of the Arctic. There is no single definition of the Arctic region.
Scientific definitions tend to emphasize climatic conditions or vegetation
zones, whereas politically guided processes, such as the Arctic Council
working groups, have used varying delimitation for deciding the southern
boundary for information to be included in the assessments. Descriptive
synonyms for the Arctic include the northern polar region, the circumpolar
north or simply the north. In this book we use these terms interchangeably
but treat only the Arctic as a proper noun for naming the region. In addition
to these circumpolar references, some countries have their own names
for designating their northernmost regions. These include the Norwegian
designation of the 'High North', the Russian 'Far North' and the 'Arctic
Zone' as well as the 'Far North' as a designation for Canada north of the
Arctic Circle. Map prepared by Hugo Ahlenius, Nordpil.

with the Arctic. The reach of such messaging—once aimed at highly specialized audiences or meant only for local communication—is not unique to the Arctic but rather an illustration of a global development in which the local can instantly become global and where no part of the world can escape the influence of global environmental and social change. Environmental scientists have in recent decades emphasized that we live in a new geological era in which human activities have a major influence on planet Earth as a system. Thus, in only a few decades, Earth system science has moved from an emphasis on the Earth as a self-regulating system with only a limited role played by human agency, captured in the Gaia metaphor (Lovelock and Margulis, 1974), to one denoted by the term Anthropocene (Crutzen, 2002; Crutzen and Stoermer, 2000). This shift in the biophysical dynamics of planet Earth is accompanied by equally fundamental social and technological changes, commonly known by the paradigmatic rubric of globalization.

Starting in the 1980s, social theory accounts of globalization have emphasized the compression of time and space through digital mediation and the dialectics of space, where *the local* can no longer be conceived without bringing in *the global*, and vice versa. Moreover, the discursive scopes of the national and the local had become too narrow to accommodate the political, economic and cultural transformations that were becoming visible. Beneath globalization were multilayered material and symbolic denationalization processes, from financial markets to economic and cultural exchanges of goods to political decision making, as well as an expansion in the technological environment through digital networks changing the media ecology. The earliest accounts of globalization in particular highlighted dynamics associated with deterritorialization as both drivers of change and the consequences of intense mediation and connectivity. Such intense spatialization through a hyper-connected media and politico-cultural environment and its societal implications are addressed in both celebratory accounts of globalization and critical research (Balibar and Wallerstein, 1991; Garnham, 1990; Mattelart, 1994; Mcchesney, 2000; Mosco, 1996; Murdock, 1993; Schiller, 1991). Hafez (2007), for one, notes the mythic role of the globalization paradigm and the way it 'discursively obfuscates . . . the local, national, and regional' (Christensen, 2013b: 2401). As David Morley (2014: 42) puts it: 'In some versions of the story of globalization, we are offered what I would characterize as an abstracted sociology of the postmodern, inhabited by an un-interrogated "we", who "nowadays" live in an undifferentiated global world'.

Spatial imaginations based on abstracted notions of globalization thus subsumed the particularities of locales and regions. They also failed to account for the dialectics of 'global spatiality', which oscillate between phases of deterritorialization and reterritorialization—the challenging of existing borders and how they limit economic, socio-cultural and political activities followed by the establishment of new borders as a result of such activities. Such dynamics bring about consolidated structures of spatiality as well as regulatory regimes that use these structures for the purposes of dominance and integration.

Specifically in the Arctic context, because relatively few people live in the region, economic globalization is often framed as a strong driver of change

(Andrew, 2014). A prominent example is the growing influence of large transnational companies in resource industries, such as the forestry, mining and oil and gas sectors (Keskitalo and Southscott, 2014). Globalization has thus become a key issue in analyses of the vulnerability of Arctic communities (e.g. Keskitalo, 2008) and in relation to adaptation (e.g. AMAP, 2017). However, globalization has also been discussed as an opportunity to break with old trajectories of national colonization, where local actors can jump scales directly to global economic and political contexts (Keskitalo and Southscott, 2014).

This book seeks to highlight how the region's growing connections to global economic and political systems combine with the shifts in the global communication landscape in ways that have engendered a new discursive and material terrain for debating the future of the region. The shift goes beyond new Arctic actors and the oscillating power relations within the region and among the global players that have been highlighted in the recent literature (e.g. Keil and Knecht, 2017; Paglia, 2016; Raspotnik, 2018; Dodds and Nuttall, 2016). While Arctic change is often narrated as a consequence of the physical impacts of climate change, such a narrative is too simplistic. Climate change, as such, is more than just a physical force affecting the region. It also comprises complex social and technological dimensions that affect both its causes, such as emissions of greenhouse gases, and its effects, where impacts and adaptation are nested within local contexts with economic, political and cultural dimensions (Nilsson et al., 2017). As is the case elsewhere in the world, climate change in the Arctic is closely intertwined with globalization in all its aspects. These include both international news and digitalized connectivity, as well as their implications for the discursive shaping of space. At the heart of the Arctic are intertwining narratives about its future(s) and relations between the global, the national, the regional and the local, not least in relation to responsibility for climate change (e.g. Dale and Kristoffersen, 2018).

Amid this complexity, simple images are attractive. They get our attention, especially when they allude to something we recognize. Media images of the Arctic are no different, and part of their power comes from how easily we can relate them to old colonial narratives about the region (Bravo and Sörlin, 2002). Examples include headlines about 'a race for resources', images related to national sovereignty and identity, and the renewed emphasis on the risk of military conflict. These and other images are also power tools, as they make it more difficult to see other perspectives. They create a frame within which we understand the region and act as an effective filter of new information, which is absorbed or ignored depending on how well it fits our preconceived notions. Media frames become especially important in discussing the future of a region that few people have first-hand experience of or deep knowledge about. At the same time, the dynamics of change themselves make it timely for many actors to position themselves in relation to what the future might bring and how they would like to either be a part of it or absolve themselves of responsibility.

Why does all this matter? A short answer would be that images and narratives have generative power in influencing the positionalities of various actors and their claims to a legitimate right to make decisions about the region's future.

The longer answer is elaborated throughout the chapters of this book and based on the notion that the Arctic of today illustrates dynamic shifts that are global in scale. We need to understand the scope of such shifts—beyond the effects of a warmer climate and beyond Internet connectivity or economic globalization—if we wish to develop political solutions that enhance human well-being rather than adding to social tensions and human insecurity. Understanding the context in which visions of the future are shaped becomes even more urgent given the call in the 2018 report of the Intergovernmental Panel on Climate Change for immediate and drastic measures to cut emissions of greenhouse gases (IPCC, 2018). Transitioning to a post-petroleum world will add yet another dimension to both global and Arctic change.

Changes in the physical geography of the Arctic play into geopolitics, in both the classical and the constructivist sense. For some Arctic states and especially those with coasts facing the Arctic Ocean—Russia, Canada, the United States, Greenland (and by extension Denmark), Norway and Iceland—the extent to which the ice recedes has consequences for industrial activities, commercial shipping and tourism, with further implications for governance. Longer periods of an ice-free Arctic also make the Arctic attractive to global players, such as China and the European Union (EU), and big corporations. The EU's interests exemplify the complex connections between changes in geography, policy choices, and its role of representing its members' and affiliated countries' interests as well as those of the corporate players with which the EU is linked in economic terms.

While recognizing the importance of how states fashion their international relations and foreign policy based on their geographic positioning (i.e. classical geopolitics), this book adopts a critical understanding—or critical geopolitics—that challenges the assumption that 'the state' and other geographical constructions are fixed entities (Dodds et al., 2013; Burkart and Christensen, 2013). We thus also turn our attention to how media and mediation influence the rapidly changing, and at times fluid, role of other actors such as NGOs, commercial interests and local communities. Moreover, we suggest that a 'geoeconomic' emphasis in geopolitics is essential in an era where economic activity can be both the reason for, and a means of, contestation and conflict in the Arctic. Increased maritime traffic can be framed in relation to potential risk for accidents or oil spills, and the need for cooperation and governance that follows, but Arctic voyages can also be understood as manifestations of capacity and power to operate in challenging polar environments.

Frames and narratives

The generative power of images and narratives lies in how they place specific information within overarching discourses and frames that can serve different interests. A focus on the context in which an event or a piece of information is placed rather than the details of the text reveals the frames that situate a specific story within a societal discourse (Nisbet, 2009; Nisbet et al., 2003). As the literature on frame theory elaborates (for reviews, see Christensen and Wormbs, 2017; Pincus and Ali, 2016), frames can be described as a cognitive mechanism that people use to grasp the most relevant information in the vast amounts of sensory input they are exposed to. They help us make sense of new information because they link it to our earlier understandings but also exclude or reinterpret information that does not comfortably fit our priorities, and thereby also set the boundaries for what becomes visible and audible. In political discourse, frames make certain debates legitimate and natural, while other ways of describing something can appear odd or out of place. The process of framing and reframing is a key aspect of policy-shaping processes as the reframing of an issue can make it more urgent or relevant for a larger group of people. For the Arctic, media and connectivity that facilitates mediation play central roles in public understanding of the region and thus in shaping narratives that can have geopolitical implications.

Mediation takes place in a range of contexts, such as the articulation of national Arctic policies, texts and images in popular culture, museum exhibits and new media, and the role of such expressions in public life and politics has come into increasing focus in studies of the critical geopolitics of the Arctic (e.g. Dittmer et al., 2011; Dittmer and Dodds, 2013; Steinberg et al., 2015; Wegge and Keil, 2018). Secondary information about the region is likely to play a prominent role in how the Arctic is framed, as relatively few people have any first-hand knowledge of the region. News media play a particular role as they have the potential to reach large audiences, either directly or indirectly, through narrowcasting and spillover to social media.

Before the late 1990s, the Arctic did not figure prominently in the mainstream media. Over the past two decades, however, various media have been paying more attention to the region, not least due to the impacts of climate change. While most studies of media discourses focus on recent years and specific issues, a study on how the Canadian press has covered the Arctic in the past 30 years shows an exceptional growth in media attention from 252 stories over the whole of the 1970s to over 1,000 stories in 2013 alone (Nicol, 2013). Other studies indicate that a substantial spike in media interest arose with the publication of the Arctic Climate Impact Assessment in 2004 (Chater and Landriault, 2016; Steinberg et al., 2014; Tjernshaugen and Bang, 2005), but even more prominent was the attention that accompanied the unexpected record Arctic Ocean sea-ice loss in 2007 (Christensen et al., 2013). Arctic climate change became a meta-event in the various media's coverage of global climate change and a context in which journalists began to tap into a range of issues, such as access to resources, new shipping opportunities, the risk of geopolitical conflict and the plight of polar bears (Christensen, 2013a). These stories were not exclusively about the Arctic as such but featured a *topical multiplicity* (ibid.) whereby the Arctic became a Christmas tree on which to hang media savvy 'hot' topics such as military conflict and resource geopolitics, at least in the western media. Many stories involved both local and global concerns, creating a sense of *scalar transcendence* (ibid.) of the local-regional Arctic being, in fact, global. Nonetheless, these changes in the coverage do not appear to have created new media space for the concerns of people living in the region, such as regional economic development, social welfare and food security, unless they fit into another overarching frame, such as global environmental change. Chapter 2 discusses coverage of the Arctic in the news media in more detail.

Narratives are particularly powerful when they can tap into earlier stories, and Arctic history is ripe with stories that project external ambitions onto the region (Bravo and Sörlin, 2002). As actor-network theory shows (Callon, 1986; Latour, 1987), a convincing storyline does not just enlarge the network of actors that share the same vision and conception of reality. If powerful enough, it can also translate into technologies and socio-technical systems with their own path dependencies. Furthermore, if the narratives and their associated networks of actors become enshrined in the norms and procedures of governance mechanisms, their power is further cemented (Avango et al., 2013). Narratives are thus part of social-ecological-technological systems that can feature system dynamics that go far beyond the narrative as such. It is therefore important to scrutinize the interplay between the discursive shaping and reshaping of the region and the impacts of the material constructions that are made possible by certain narratives. Critically important for our interest, these include communications infrastructure, such as satellites and fibre optic technologies, as well as the range of software algorithms and services that create platforms for sharing information, values and interests. Such materialities have profound impacts on the mobility of narratives, how they interact with other storylines and their potential influence on decisions that shape the future.

Narratives do not appear out of nowhere, however. At one level they have their basis in specific interests, but interests start to become narratives more concretely when they are expressed, and as speech or other actions they can travel further. Putting ideas into words or images is an action that is often aimed at creating change. When words and images feed into narratives that are further mediated, they can have the power to shape overarching narratives and change them in ways that can ultimately have geopolitical consequences, as is illustrated for the Arctic in Chapters 3 and 4.

Shifting media ecosystems

The media, in all its discursive shades and materialities, is a big player in the Anthropocene. At the same time, the news media is facing many challenges in a cultural and financial landscape of economic downturns and consolidations, leading to the downsizing of the environmental desks of 'media giants' such as globally prominent newspapers. Furthermore, these media outlets are affected by social media filter bubbles, which influence news reporting and editorial choices. Questions therefore arise regarding 'metacommunication', or the 'communication of communication', and how the current landscape might influence the dominant discourses that circulate and, in the longer run, broader political decision making on the Arctic.

A combination of theoretical perspectives is needed to address the issue of the interconnected factors that shape news and media reporting on anthropogenic environmental change in general, and Arctic change in particular. Two areas provide a starting point for a more complex understanding of the relationship between the so-called legacy media (traditional outlets), online platforms such as social media and offline forms of communication such as art, music and orality: disintermediation and media ecology.

First, disintermediation is related to the extent to which mainstream, 'traditional' or 'legacy' media maintain or lose power and relevance in an era of citizen, activist and social media. As Aday et al. (2013) note, a core of researchers insists that the power of these elite gatekeepers has diminished, and that the increase in horizontal creation and sharing of information will disintermediate legacy media by eroding their near-monopoly position as a bridge between citizens, and between citizens and the state. An alternate view is that this disintermediation is not taking place, and that major media corporations have maintained their ability to frame events and set the public agenda. It is clear that the rise of social media platforms over the past 15 years has given environmental activists and indigenous peoples and others new avenues for reaching out to large numbers of individuals without the need to rely upon mainstream media coverage for exposure. Yet, the fundamental question is whether or not this by-passing of traditional outlets is an exception rather than the norm. To put it another way: *to what extent* does self-produced activist or on-site content, released via social media platforms, still rely upon pick-up by mainstream news outlets for the level of major exposure needed to make a real impact? And, in addition, to what extent does mainstream media exposure impart a degree of (editorial) *gravitas* and credibility to material produced by, for example, activists?

In order to avoid the 'zero-sum-game' of seeing media influence as an all-or-nothing contest between a limited number of media actors—primarily mainstream media news outlets and social media platforms—media ecology can be a useful theoretical framework (see Cottle, 2011; Robertson, 2013; Tufekci and Wilson, 2012). The conception of media ecology is rooted in the work of Scolari (2012, 2013), and his notion of an 'intermedia' variant of ecology (as opposed to an 'environmental' conception), within which different media forms co-exist, as plants and animals co-exist within a classical ecosystem. For Scolari, the key is not to examine each media form in isolation, but to consider the relationships between media (2013: 1419) and how various media forms (electronic and non-electronic) influence each other. In relation to the coverage of the Arctic, a media ecology perspective would take into account the domination of storytelling power. Lessons from other parts of the world, for example, show that mainstream media rarely make use of indigenous perspectives, and that, 'indigenous actors have limited access to the types of communication required to reach a larger audience' (Graf, 2016: 10). Roosvall and Tegelberg (2016) found that when asked about their understanding of the media ecosystem (national, mainstream, local, alternative and non-indigenous) within which they operate, indigenous activists found mainstream outlets to be exclusive, but so powerful that they could not be ignored. The authors note that the 'media logic in the mainstream realm of the news ecology is strongly connected to power' and that, for activists, 'it is necessary to connect to this power if a message is to be widely heard and respected' (Roosvall and Tegelberg, 2016: 98). Crucially, respondents also noted the underlying political economy of media systems, highlighting 'the need for a re-balancing of current media ecologies—which unlike natural ecologies are constructed ecologies closely connected to distribution of means' (p. 98). The more pessimistic view of political economy is that attempts by indigenous populations to get their issues raised in mainstream outlets are ultimately futile because the demands of private capital will always side with 'industry and governments rather than the people' (Roosvall and Tegelberg, 2016: 95).

Scale, politics and power

For a region that has not traditionally had a strong 'own voice' in the media and where political priorities have most often been determined by actors outside the region, the questions of narrative power, framing and mediated discourse are particularly significant. As is discussed in Chapter 2, media discourses about the region vary between different countries and also over time. For example, a study of Canadian press coverage (Nicol, 2013) showed that discourses related to the economy and development once dominated coverage of the Arctic region, but that in recent decades these have been supplemented by increasing attention to science and the environment. Of specific interest to the overarching question addressed in this book—the potential for and limitations of international regional governance—is how discourses relate to scalar preferences and how such preferences might play out in the politics of Arctic change. Issues related to scalar preferences and politics have been treated extensively in the literature, from political science and geography to media studies (Brenner, 2001; Brown and Purcell, 2005; Couldry and McCarthy, 2004; Dittmer and Dodds, 2008, 2013; McCarthy, 2005; Smith, 1992; Swyngedouw, 2000, to list just a few). A key aspect is that they can affect preferred levels of governance (Lebel et al., 2005; e.g. Swyngedouw, 1997), in that a specific scalar framing can make it very difficult to think about an issue as anything other than local, national or global, depending on how the discourse has developed. Framing the Arctic as a global interest invites different actors into the discussion than a framing that emphasizes the local or the national would. Framings, therefore, conserve certain power relations, in a similar way to how established governance structures favour those who already have a seat and disfavour those whose interests do not fit nicely into the established frame. The power of framing is thus ultimately about power over decision making.

Scalar framings also serve to assign responsibility. For example, in discussing climate change, responsibility for reducing emissions of greenhouse gases is most often highlighted as a global issue, while adaptation is mainly seen as a local or national concern. Any deviation from this 'natural' division requires extensive political negotiations backed up by arguments that are strong enough to justify a deviation from the norm.

The transformation of the Arctic into a recognized international region in the 1990s is another example of how relevant scale is negotiated and highlights the tensions between the national as the most natural framing and the regional as the more relevant (for descriptions of this negotiation process, see Young, 1998; Tennberg, 1998; Keskitalo, 2004; for a discussion of the reframing process from a national to a regional focus in political logic, see Heininen and Nicol, 2007). These negotiations were based on the notion of the national as the highest level of political decision making. This is a scalar preference that has been naturalized to the extent that it can be difficult to think about international society in any other way and is the underlying assumption of realist international relations theories (Dunne and Schmidt, 2001). However, in the late 1980s and early 1990s, access to the region's offshore resources required peaceful cooperation, as did any attempt to address the transboundary environmental problems that were becoming apparent. This situation, along with the emergence of indigenous peoples as new transnational political actors, opened up space for a renegotiation of scalar preferences regarding the Arctic, as is discussed further in Chapter 3. Heininen and Nicol (2007) describe the emergence of a circumpolar regional perspective as a process of 'reterritorialization' within the circumpolar north and highlight how 'new voices and new discourses created a new discourse that operated at a circumpolar scale' (p. 135). It is easy to forget that writing 'Arctic' with a capital A is a relatively new phenomenon, and that 'arctic' for a long time referred mainly to northern ecosystems and landscapes without any political significance attributed to the term in an international context. Whether it is relevant to treat the Arctic as a region is still contested. For example, Keskitalo et al. (2013) argue that such a framing has imposed images and understandings based on the situation in North America on northern Fennoscandia without taking into consideration that national policy, populations and economic structures are very different, leading to a risk of 'fictionalizing the region'. According to the authors' line of reasoning (ibid.), the unique history and institutional structures of the Fennoscandia north should be understood in their own right by de-emphasizing the regional framing. However, such a de-emphasis of the regional would also place less focus on circumpolar social and environmental processes and commonalities that can have political implications—one example being whether it should be considered relevant to highlight similarities related to histories of colonization.

Each scalar perspective has its own advantages, depending on the question in focus. It is no coincidence that certain scientific disciplines have developed a strong preference for the global scale—climate science being a case in point— while others focus on local processes. The latter include ecological studies, where local fieldwork to understand ecosystem dynamics has a strong tradition. The problem in relation to power over decision making arises when one scalar preference comes to exclude other possible perspectives, or when one specific perspective is deemed scientifically correct and others not. The latter lurks in the background in the use of the term 'fit', which has been used in the study of how institutions affect the relationship between societies and the environment. Embedded in the notion of fit is the understanding that there is an ideal scope and scale of governance (Folke et al., 1998, 2007; Galaz et al., 2008; Young, 2002). Most often, a sort of ecosystem process has been considered the relevant starting point for defining fit, but other spatial logics also appear in the literature, such as social fit and social-ecological system fit (Epstein et al., 2015).

Following arguments articulated by Clement (2012) and insights from critical studies of scale, we see a need to go beyond identifying the 'right' institutional fit to instead focus on how power and discourses frame what is considered an appropriate scalar delimitation. By recognizing and focusing analytical efforts on how any framing of fit is negotiated, in policy messages as well as in the media, it becomes possible to reveal interests, perspectives and shifts in power relations that might not be equally obvious within an established discourse.

Notions of appropriate scale of governance have shifted substantially over time. For example, while many local societies have developed systems for managing common pool resources (Ostrom, 1990), states have increasingly taken on the role of politically legitimate administrative power, even if some decisions are then delegated to subnational governments. During the 20th century, particularly in the aftermath of World War II, this national scale was challenged when human security as a collective interest became an issue of global politics. One example is when food insecurity became framed as a question on which international collaboration was necessary (Mayne, 1947). The global regulation of trade, which later developed into the World Trade Organization, also stems from the post-World War II era. A few decades later, when pollution and the risk of depletion of resources became hotly debated political topics (Worster, 1994), it also became clear that national borders were more porous than political theorists had previously imagined. Following publication of the Club of Rome report, *Limits to Growth*, in 1972, a global systems perspective entered the mainstream political debate (Meadows et al., 1972). In the same year, the United Nations held its first global summit on the human environment in Stockholm, on the theme 'Only One Earth' (Ward and Dubos, 1972).

The focus in political discussion about the relationship between people and the natural environment has thus moved from the local to the global, even if there are also examples of a counter-movement back to stronger local influence, especially in cases where top-down management has failed. The pre-eminence of the global framing has its roots in the images of planet Earth from space, along with mathematical models of Earth as a single system that were enabled by computer technologies (Miller and Edwards, 2001) and concerns over the limits to the global resource supply (Meadows et al., 1972). The global scale perspective also made inroads into the social sciences in the 1970s, not least with world systems theory and its critique of the nation state as the only unit of analysis, and focus on core–periphery economic and power relations (Wallerstein, 1974). In the 1990s, mainstream political scientists started to highlight the complex interdependence of international systems (Keohane and Nye, 1994). In the growing plethora of international environmental agreements, political scientists turned their attention to the dynamics of global governance as a system that went beyond earlier notions of international society as anarchic (Young, 1997).

Despite its relatively recent history and the critique of reductionist understandings of globalization, the persistence of the unreflective uses of *the global* is a good example of the power of framing that has implications for one of the major challenges facing the Arctic region—climate change. In the case of climate change mitigation, a global framing has been used as a rhetorical tool to avoid taking responsibility locally, nationally and regionally, as is discussed in Chapter 5. The point here is not to question the legitimacy of the global framing as a useful analytical starting point for understanding biophysical and social processes that defy national borders, or to question efforts to create governance mechanisms that go beyond the national. However, it becomes problematic when one scalar perspective becomes so dominant that it can be used to preclude political action in all other scalar contexts. Why, for example, have the Arctic nations, which all highlight the challenges related to the severe impacts of climate change in their national policies, not made the Arctic region a leader in reducing reliance on the use and production of fossil fuels? The answer, of course, lies in what is politically palatable, albeit not according to any 'natural law', that regions are not *fit* to take a lead in such a transformation. Indeed, as is discussed further in Chapter 5, this relates to the strong geopolitical interests connected with the production of oil and gas in the Arctic, which it might be convenient to conceal beneath another narrative—that climate change is a common global responsibility. A highly dominant scalar perspective can also hide important information about the world that would be more visible from other scalar perspectives—information that might be critical to achieving specific goals, such as securing and enhancing human well-being. For example, research on adaptation to climate change inspired by an anthropological focus on *the local* has shown that shifts in climate are not the only drivers of change locally, and that local and regional histories and dynamics play an important role in the capacity of societies to deal with global drivers of change, including shifts in climate conditions (Nilsson et al., 2017).

On the Arctic, a global framing often dominates the discourse. This is no surprise given that it is the region's global significance that has made the Arctic relevant to media consumers as well as political actors beyond the region. A catchphrase used by scientists and politicians alike is that what happens in the Arctic does not stay in the Arctic. Efforts have also been devoted to establishing the phrase 'Global Arctic' (Heininen and Finger, 2017). While some media imagery depicts local events, including polar bears on ice floes and indigenous people in the context of their traditional livelihoods, the aim is often to emphasize a global issue, especially climate change, and to use the Arctic as a tool for raising global political awareness.

Scalar perspectives have political implications. It is no coincidence that the European Parliament in 2008 articulated the idea of an Arctic Treaty to govern the circumpolar north, modelled on the Antarctic Treaty (European Parliament, 2008). This would give the EU, as a supranational actor, a legitimate role that it currently lacks in shaping the region's future by way of putting the emphasis on territorial presence and the connected sovereign rights. The view of the southern polar region as the common heritage of mankind is another example of global framing, which stems from the global research initiatives of the International Geophysical Year 1957–58, that actors have since attempted to apply to the Arctic. However, the idea of an Arctic Treaty led to immediate pushback, not least from states that already have well-established territory in the circumpolar north and indigenous peoples who saw a potential Arctic Treaty as threat to their right to self-determination. Instead, Arctic states began to emphasize existing mechanisms for international cooperation, such as the Arctic Council and the UN Convention on the Law of the Sea (UNCLOS). Incidentally, these are governance mechanisms in which certain states—and in the case of the Arctic Council also indigenous peoples' organizations—have a privileged position in comparison with other international actors. The eight countries that are represented on the Arctic Council also began to speak about themselves as the 'Arctic nations' in newly issued Arctic policies. This stance was countered by various imaginative expressions from other actors seeking to challenge any claims to exclusive privileges, including China's claim to be a 'near-Arctic state' and language in its Arctic policy that frames the region as of common global interest (Xinhua, 2018).

An Arctic regional framing has been in the making since the 1990s and the establishment of Arctic regional cooperation (Keskitalo, 2004). However, from the perspective of national high-level priorities, this development appears to have taken place mostly below the political radar until the actual and potential future impacts of climate change started to become apparent, particularly following the 2007 Arctic Ocean sea-ice minimum. It suddenly became important to reassert the national scale, in contrast to framing the Arctic region as a global concern. The rationale is obvious: it defines the actors that have a legitimate right to make decisions about the region. The relative roles of the regional scale perspective and national framing has been a recurring theme in negotiating circumpolar cooperation, where the Arctic Council is still a high-level forum and not governed by legally binding obligations. The United States, for one, has been adamant that it should not take on any decision-making rights from the national level (Bloom, 1999; National Security Presidential Directive 66 on Arctic Region Policy, 2009). This lack of a regional perspective in power over decision making is also visible in the fact that the legally binding agreements that have been signed under the auspices of the Arctic Council have a focus on coordination among the Arctic states rather than creating another level of governance that would supersede the power of states. This can be contrasted with the state of affairs in the European Union, where the member states have given up some of their rights to the supranational level of decision making in the EU's administrative and political machinery.

The regional of today; the regional of tomorrow?

While the Arctic appears in the title of this book, the argument put forward is not just about the Arctic but also a more general question about the role of the regional international scale and its relevance to governance. This question is especially pertinent when the recognized challenges are global in scope while global governance initiatives struggle to make a difference, the need to reduce emissions of greenhouse gases to halt climate change being a prime example. Moreover, the positive expectations of globalization are increasingly being questioned and a return to a focus on nation states rather than international and transnational responsibilities is gaining political momentum. Can international regions provide a middle ground in which international collaboration can still make a difference? Is regional cooperation between fewer actors more manageable than trying to negotiate among the 193 member states of the United Nations?

At first sight, it would appear so, but a close study of a region such as the Arctic could provide an empirical foundation for highlighting the strengths and shortcomings in relation to the ambitions and priorities of various actors. From a geopolitical point of view, the Arctic is an especially interesting case for two reasons. First, cooperation has continued despite the major rifts between Russia and the rest following the annexation of Crimea. Second, and linked to climate change and energy geopolitics, its vast reserves of oil and gas mean that international agreements among the Arctic states could play a central role in one of today's major political challenges—the need to transition to a society that does not depend of fossil fuels for its energy needs. The challenges are huge, not least that of acquiring the political will. Russia is the largest Arctic country by both population and geographical extent and much of Russia's Arctic population is dependent on industry and the extractive sector for its employment and livelihood. In addition, industrial activity in the Russian Arctic amounts to over 11 per cent of the country's gross domestic product. It is a similar story, albeit not to the same extent, for the population of Alaska in the USA, many of whom regard climate change and energy-related science with scepticism. This points to the delicate equilibrium between high politics and questions of livelihood in the region, as well as the role of mediation in communicating alternative perspectives and solutions. We return to these questions in Chapter 5.

The theoretical appeal lies in the fact that the study offers a window on how international society functions today—a world of nested and interacting social, ecological, economic and technological systems, where the pervasiveness of the media and mediations play a central role in shaping perceptions of current challenges and available opportunities. While there are many issues of common concern, one stands out in particular: the need to find a path to a sustainable post-petroleum society. Past emissions and the momentum of the climate system mean that climate change cannot be eliminated, but there is still time to affect its pace and eventual magnitude as well as its impacts on society. Access to energy has played a central role in human history, and shifts in energy regimes have had profound consequences for all aspects of society (McNeill, 2001). A deliberate shift away from the current fossil fuel-based energy regime is not welcomed by everyone. As noted above, in the Arctic the stakes are high, with strong economic and political interests to the fore. It is thus a useful case in which to highlight the potential for a regional international regime to overcome some of the inevitable inertia caused by existing power geometries. One way to overcome inertia in an existing regime is a major shift in the distribution of power among the actors involved. A more likely scenario would be a reframing of interests in ways that make change palatable within existing power structures.

Organization of the book

Following this introductory chapter, Chapter 2 is an in-depth examination of the content-related and infrastructural aspects of media and mediation. Chapter 3 discusses how the Arctic came to be narrated as a region in the 1980s. This is followed by a discussion of more contemporary developments in Chapter 4. Chapter 5 has a specific focus on media and policy narratives in relation to the potential for a post-petroleum future. Chapter 6 concludes the book with a discussion on the implications of the discursive and material dynamics treated in the earlier chapters and returns to the generic question of the role of regional international governance in a world where the legitimacy of democratic decision making requires attention to social, ecosystem and technological dynamics at all scales, from the local to the global.

References

Aday S, Farrell H, Freelon D et al. (2013) Watching from afar: Media consumption patterns around the Arab Spring. *American Behavioral Scientist* 57(7): 899–919. DOI: 10.1177/0002764213479373.

AMAP (2017) *Adaptation Actions for a Changing Arctic: Perspectives from the Barents Area*. Oslo: Arctic Monitoring and Assessment Programme (AMAP).

Andrew R (2014) *Socio-Economic Drivers of Change in the Arctic*. Oslo: Arctic Monitoring and Assessment Programme.

Avango D, Nilsson AE and Roberts P (2013) Arctic futures: Voices, resources and governance. *Polar Journal* 3(2): 431–446. DOI: https://doi.org/10.1080/21548 96X.2013.790197.

Balibar E and Wallerstein IM (1991) *Race, Nation, Class: Ambiguous Identities*. London and New York: Verso.

Bloom E (1999) Establishment of the Arctic Council. *The American Journal of International Law* 93(3): 712–722.

Bravo MT and Sörlin S (eds) (2002) *Narrating the Arctic: A Cultural History of Nordic Scientific Practices*. Canton, MA: Science History Publications.

Brenner N (2001) The limits to scale? Methodological reflections on scalar structuration. *Progress in Human Geography* 25(4): 591–614. DOI: 10.1191/030913201682688959.

Burkart P and Christensen M (2013) Geopolitics and the popular. *Popular Communication* 11(1): 3–6. DOI: 10.1080/15405702.2013.751853.

Brown J and Purcell M (2005) There's nothing inherent about scale: Political ecology, the local trap, and the politics of development in the Brazilian Amazon. *Geoforum* 36(5): 607–624. DOI: 10.1016/j.geoforum.2004.09.001.

Callon M (1986) The sociology of an actor-network: The case of the electric vehicle. In: *Mapping the Dynamics of Science and Technology*. London: Macmillan Press, pp. 19–34.

Chater A and Landriault M (2016) Understanding media perceptions of the Arctic Council. In: Heininen L, Exner-Pirot H, and Plouffe J (eds) *Arctic Yearbook 2016*. Akureyri, Iceland: Northern Research Forum, pp. 61–74. Available at: https://arcticyearbook. com/arctic-yearbook/2016/2016-scholarly-papers/168-understanding-media-perceptions-of-the-arctic-council (accessed 13 March 2019).

Christensen M (2013a) Arctic climate change and the media: The news story that *was*. In: Christensen M, Nilsson AE and Wormbs N (eds) *Media and the Politics of Arctic Climate Change: When the Ice Breaks*. New York: Palgrave Macmillan, pp. 26–51.

Christensen M (2013b) *Trans*National media flows: Some key questions and debates. *International Journal of Communication* 7: 2400–2418.

Christensen M and Wormbs N (2017) Global climate talks from failure to cooperation and hope: Swedish news framings of COP15 and COP21. *Environmental Communication* 11(5): 682–699. DOI: 10.1080/17524032.2017.1333964.

Christensen M, Nilsson AE and Wormbs N (eds) (2013) *Media and the Politics of Arctic Climate Change: When the Ice Breaks*. New York: Palgrave Macmillan.

Clement F (2012) For critical social-ecological system studies: Integrating power and discourses to move beyond the right institutional fit. *Environmental Conservation* 40(1): 1–4. DOI: http://dx.doi.org/10.1017/S0376892912000276.

Cottle S (2011) Media and the Arab uprisings of 2011: Research notes. *Journalism: Theory, Practice & Criticism* 12(5): 647–659. DOI: 10.1177/1464884911410017.

Couldry N and McCarthy A (2004) *MediaSpace: Place, Scale and Culture in a Media Age*. London: Routledge.

Crutzen PJ (2002) The 'Anthropocene'. *Journal de Physique IV (Proceedings)* 12(10): 1–5. DOI: 10.1051/jp4:20020447.

Crutzen PJ and Stoermer EF (2000) The 'Anthropocene'. *Global Change Newsletter* (41): 17–18.

Dale B and Kristoffersen B (2018) Post-petroleum security in a changing Arctic: Narratives and trajectories towards viable futures. *Arctic Review on Law and Politics* 9: 244–261. DOI: 10.23865/arctic.v9.1251.

Declaration of Arctic Mayors (2017) Arctic Portal. Available at: https://arcticportal.org/ap-library/news/1909-declaration-of-arctic-mayors (accessed 18 January 2018).

Dittmer J and Dodds K (2008) Popular geopolitics past and future: Fandom, identities and audiences. *Geopolitics* 13(3): 437–457. DOI: 10.1080/14650040802203687.

Dittmer J and Dodds K (2013) The geopolitical audience: Watching Quantum of Solace (2008) in London. *Popular Communication* 11(1): 76–91. DOI: 10.1080/15405702.2013.747938.

Dittmer J, Moisio S, Ingram A et al. (2011) Have you heard the one about the disappearing ice? Recasting Arctic geopolitics. *Political Geography* 30(4): 202–214. DOI: 10.1016/j.polgeo.2011.04.002.

Dodds K and Nuttall M (2016) *The Scramble for the Poles: The Geopolitics of the Arctic and Antarctic*. Cambridge, UK; Malden, MA: Polity Press.

Dodds K, Kuus M and Sharp JP (eds) (2013) *The Ashgate Research Companion to Critical Geopolitics*. Farnham and Burlington, VT: Ashgate.

Dunne T and Schmidt BC (2001) Realism. In: *The Globalization of World Politics: An Introduction to International Relations*. Oxford: Oxford University Press, pp. 141–161.

Epstein G, Pittman J, Alexander SM et al. (2015) Institutional fit and the sustainability of social–ecological systems. *Current Opinion in Environmental Sustainability* 14. Open Issue: 34–40. DOI: 10.1016/j.cosust.2015.03.005.

European Parliament (2008) Arctic Governance: Texts adopted – Thursday, 9 October 2008 – P6_TA(2008)0474. European Parliament. Available at: www.europarl.europa.eu/sides/getDoc.do?type=TA&reference=P6-TA-2008-0474&language=EN (accessed 18 January 2018).

Folke C, Pritchard L, Berkes F et al. (1998) *The Problem of Fit Between Ecosystems and Institutions*. No 2. Bonn: International Human Dimensions Programme.

Folke C, Pritchard L, Berkes F et al. (2007) The problem of fit between ecosystems and institutions: Ten years later. *Ecology and Society* 12(1): 30.

Galaz V, Olsson P, Hahn T et al. (2008) The problem of fit among biophysical systems, environmental and resource regimes, and broader governance systems: Insights and emerging challenges. In: *Institutions and Environmental Change: Principal Findings, Applications, and Research Frontiers*. Cambridge, MA: MIT Press, pp. 147–186.

Garnham PN (1990) *Capitalism and Communication: Global Culture and the Economics of Information*. London, Newbury Park: SAGE Publications.

Graf H (ed.) (2016) *The Environment in the Age of the Internet: Activists, Communication, and the Digital Landscape*. Cambridge: Open Book Publishers.

Hafez K (2007) *The Myth of Media Globalization*. Cambridge: Polity Press.

Heininen L and Finger M (2017) The 'Global Arctic' as a new geopolitical context and method. *Journal of Borderlands Studies* 33(2): 199–202. DOI: 10.1080/08865655.2017.1315605.

Heininen L and Nicol H (2007) The importance of Northern Dimensions foreign policies in the geopolitics of the circumpolar North. *Geopolitics* 12(1): 133–165. DOI: 10.1080/14650040601031206.

IPCC (2018) Global Warming of 1.5°C. An IPCC Special Report on the impacts of global warming of 1.5°C above pre-industrial levels and related global greenhouse gas emission pathways, in the context of strengthening the global response to the threat of climate change, sustainable development, and efforts to eradicate poverty [V Masson-Delmotte, P Zhai, HO Pörtner, D Roberts, J Skea, PR Shukla, A Pirani, W Moufouma-Okia, C Péan, R Pidcock, S Connors, JBR Matthews, Y Chen, X Zhou, MI Gomis, E Lonnoy, T Maycock, M Tignor, T Waterfield (eds.)]. Available at: www.ipcc.ch/report/sr15/ (accessed 23 October 2018).

Keil K and Knecht S (eds) (2017) *Governing Arctic Change*. London: Palgrave Macmillan UK. DOI: 10.1057/978-1-137-50884-3.

Keohane RO and Nye JS (1994) *Power and Interdependence*. New York: HarperCollins.

Keskitalo ECH (2004) *Negotiating the Arctic: The Construction of an International Region*. London: Routledge.

Keskitalo ECH (2008) *Climate Change and Globalization in the Arctic: An Integrated Approach to Vulnerability Assessment*. London: Earthscan.

Keskitalo ECH and Southscott C (2014) Globalization. In: Larsen JN and Fondahl G (eds) *Arctic Human Development Report: Regional Processes and Global Challenges*. TemaNord 2014:567. Copenhagen: Nordic Council of Ministers, pp. 398–421.

Keskitalo ECH, Malmberg G, Westin K et al. (2013) Contrasting Arctic and mainstream Swedish descriptions of northern Sweden: The view from established domestic research. *Arctic* 66(3): 351–365.

Kristoffersen C (2017) Historic Arctic mayoral declaration. *High North News*, 23 May. Available at: www.highnorthnews.com/op-ed-historic-arctic-mayoral-declaration/ (accessed 18 January 2018).

Latour B (1987) *Science in Action*. Cambridge, MA: Harvard University Press.

Lebel L, Garden P and Imamura M (2005) The politics of scale, position, and place in the governance of water resources in the Mekong region. *Ecology and Society* 10(2): 18.

Lovelock JE and Margulis L (1974) Atmospheric homeostasis by and for the biosphere: The Gaia hypothesis. *Tellus A* 26(1–2). DOI: 10.3402/tellusa.v26i1-2.9731.

Mattelart A (1994) *Mapping World Communication: War, Progress, Culture*. Minneapolis: University of Minnesota Press.

Mayne JB (1947) FAO–The background. *Review of Marketing and Agricultural Economics* 15(10): 354–384.

McCarthy J (2005) Scale, sovereignty, and strategy in environmental governance. *Antipode* 37(4): 731–753. DOI: 10.1111/j.0066-4812.2005.00523.x.

Mcchesney RW (2000) The political economy of communication and the future of the field. *Media, Culture & Society* 22(1): 109–116. DOI: 10.1177/016344300022001006.

McNeill JR (2001) *Something New Under the Sun: An Environmental History of the Twentieth-Century World*. London: Penguin.

Meadows DH, Meadows DL, Randers J et al. (1972) *The Limits to Growth: A Report for the Club of Rome's Project on the Predicament of Mankind*. London: Earth Island.

Miller CA and Edwards PN (eds) (2001) *Changing the Atmosphere: Expert Knowledge and Environmental Governance*. Cambridge, MA: MIT Press.

Morley D (2014) On living in a techno-globalized world: Questions of history and geography. In: Servaes J (ed.) *Technological Determinism and Social Change: Communication in a Tech-Mad World*. London: Lexington Books, pp. 41–50.

Mosco V (1996) Myths along the information highway. *Peace Review* 8(1): 119–125. DOI: 10.1080/10402659608425939.

Murdock G (1993) Communications and the constitution of modernity. *Media, Culture & Society* 15(4): 521–539. DOI: 10.1177/016344393015004002.

National Security Presidential Directive 66 on Arctic Region Policy (2009) Available at: https://fas.org/irp/offdocs/nspd/nspd-66.htm (accessed 6 October 2017).

Nicol H (2013) Natural news, state discourses & the Canadian Arctic. In: Heininen L, Exner-Pirot H and Plouffe J (eds) *Arctic Yearbook 2013*. Northern Research Forum, pp. 211–236.

Nilsson AE, Hovelsrud GK and Karlsson M (2017) Synthesis. In: *Adaptation Action for a Changing Arctic Barents Regional Report*. Oslo: Arctic Monitoring and Assessment Programme (AMAP), pp. 253–266.

Nisbet MC (2009) Communicating climate change: Why frames matter for public engagement. *Environment: Science and Policy for Sustainable Development* 51(2): 12–23. DOI: 10.3200/ENVT.51.2.12-23.

Nisbet MC, Brossard D and Kroepsch A (2003) Framing science: The stem cell controversy in an age of press/politics. *Harvard International Journal of Press/Politics* 8(2): 36–70.

Ostrom E (1990) *Governing the Commons: The Evolution of Institutions for Collective Action*. Cambridge: Cambridge University Press.

Paglia E (2016) The telecoupled Arctic: Ny-Ålesund, Svalbard as scientific and geopolitical node. In: *The Northward Course of the Anthropocene: Transformation, Temporality and Telecoupling in a Time of Environmental Crisis*. Stockholm: KTH Royal Institute of Technology. PhD Dissertation KTH Royal Institute of Technology, Division of History of science, technology and environment, pp. 17–29. Available at: http://kth.diva-portal.org/smash/get/diva2:881415/FULLTEXT02.pdf.

Pincus R and Ali SH (2016) Have you been to 'The Arctic'? Frame theory and the role of media coverage in shaping Arctic discourse. *Polar Geography* 39(2): 83–97. DOI: 10.1080/1088937X.2016.1184722.

Raspotnik A (2018) *The European Union and the Geopolitics of the Arctic*. Cheltenham, UK; Northampton, MA: E E Elgar. Available at: www.e-elgar.com/shop/the-european-union-and-the-geopolitics-of-the-arctic (accessed 20 June 2018).

Robertson A (2013) Connecting in crisis: 'Old' and 'new' media and the Arab Spring. *International Journal of Press/Politics* 18(3): 325–341. DOI: 10.1177/1940161213484971.

Roosvall A and Tegelberg M (2016) Natural ecology meets media ecology: Indigenous climate change activists' views on nature and media. In: Graf H (ed.) *The Environment in the Age of the Internet*. Cambridge: Open Book Publishers, pp. 75–104. Available at: www.openbookpublishers.com/product/484 (accessed 7 March 2019).

Schiller HI (1991) Not yet the post-imperialist era. *Critical Studies in Mass Communication* 8(1): 13–28. DOI: 10.1080/15295039109366777.

Scolari CA (2012) Media ecology: Exploring the metaphor to expand the theory. *Communication Theory* 22(2): 204–225. DOI: 10.1111/j.1468-2885.2012.01404.x.

Scolari CA (2013) Media evolution: Emergence, dominance, survival, and extinction in the media ecology. Available at: http://repositori.upf.edu/handle/10230/26010 (accessed 12 October 2018).

Smith N (1992) Geography, difference and the politics of scale. In: Doherty J, Graham E and Malek M (eds) *Postmodernism and the Social Sciences*. London: Palgrave Macmillan UK, pp. 57–79.

Steinberg PE, Bruun JM and Medby IA (2014) Covering Kiruna: A natural experiment in Arctic awareness. *Polar Geography* 37(4): 273–297. DOI: 10.1080/1088937X.2014. 978409.

Steinberg PE, Tasch J and Gerhardt H (2015) *Contesting the Arctic: Rethinking Politics in the Circumpolar North*. London: I B Tauris & Co Ltd.

Swyngedouw E (1997) Neither global nor local: 'Glocalization' and the politics of scale. In: Cox K (ed.) *Spaces of Globalization: Reasserting the Power of the Local*. New York: Guilford Press, Chapter 6.

Swyngedouw E (2000) Authoritarian governance, power, and the politics of rescaling. *Environment and Planning D: Society and Space* 18(1): 63–76. DOI: 10.1068/d9s.

Tennberg M (1998) *Arctic Environmental Cooperation: A Study in Governmentality*. Ashgate Publishing Company.

Tjernshaugen A and Bang G (2005) *ACIA og IPCC en sammenligning av mottakelsen i amerikansk offentlighet*. No. 2005:4. Oslo: Cicero, Center for International Climate and Environmental Research. Available at: https://brage.bibsys.no/xmlui/handle/11250/191998 (accessed 7 March 2019).

Tufekci Z and Wilson C (2012) Social media and the decision to participate in political protest: Observations from Tahrir Square. *Journal of Communication* 62(2): 363–379. DOI: 10.1111/j.1460-2466.2012.01629.x.

Wallerstein I (1974) *The Modern World-System I: Capitalist Agriculture and the Origins of the European World-Economy in the Sixteenth Century*. New York and London: Academic Press. Available at: www.jstor.org/stable/10.1525/j.ctt1pnrj9 (accessed 18 January 2018).

Ward B and Dubos R (1972) *Only One Earth: The Care and Maintenance of a Small Planet*. New York: Norton.

Wegge N and Keil K (2018) Between classical and critical geopolitics in a changing Arctic. *Polar Geography* 41(2): 87–106. DOI: 10.1080/1088937X.2018.1455755.

Worster D (1994) *Nature's Economy: A History of Ecological Ideas*. Cambridge: Cambridge University Press.

Xinhua (2018) China's Arctic Policy. Available at: www.chinadailyhk.com/articles/188/159/234/1516941033919.html (accessed 19 February 2018).

Young OR (1997) Rights, rules, and resources in world affairs. In: *Global Governance*. Cambridge, MA: MIT Press, pp. 1–23.

Young OR (1998) *Creating Regimes: Arctic Accords and International Governance*. Ithaca, NY: Cornell University Press.

Young OR (2002) *The Institutional Dimensions of Environmental Change: Fit, Interplay, and Scale*. Cambridge, MA: MIT Press.

2   Media narratives–media cartographies

Throughout the 20th and into the 21st century, the Arctic has been inscribed with a diverse range of political and popular imaginings (Steinberg et al., 2015), which have added new layers to the colonization narratives of the 18th and 19th centuries (Bravo and Sörlin, 2002). Their mediation reflects shifts in the social landscape over time. Early depictions of polar explorers, missionaries and treasure hunters in literature and paintings have given way to present-day social media messaging and activist performances featuring images of melting ice and starving polar bears. Depictions of Arctic geopolitics in artistic and news representations, feature films and literature which formerly focused on the potential deployment of nuclear weapons and related hostilities have expanded to include imaginings of peace and cooperation, such as Gorbachev's widely mediated Murmansk initiative in 1987 and photographs of Arctic Council events with the flags of the eight Arctic countries and the six indigenous peoples' organizations.

While many Arctic imaginaries persist over time, shifts in geopolitics as well as in modes of mediation and media cartography create new contexts. It is notable how the availability of satellite measurements combined with other sophisticated methods of ice analysis have provided data and insights about a changing Arctic Ocean that sets the past three and a half decades apart from any preceding era. Submarine measurements between 1987 and the mid-1990s showed that the ice cover became unusually thin in the mid-1990s. Further observational records that combined remote sensing, submarine data, modelling and field and marine expeditions underlined the variability of the Arctic atmosphere, its ocean and ice cover (Rothrock et al., 2003). This information has been mediated with powerful imagery that provides a gaze on the region from a circumpolar perspective but in a global context (Wormbs, 2013). Already in 1995, the Intergovernmental Panel on Climate Change (IPCC) was noting that there was evidence of 'discernible human influence on global climate' (IPCC, 1995). Later IPCC reports, particularly in the 2000s, increasingly underlined the scientific consensus on anthropogenic climate change. Meanwhile, the role of the Arctic in this change became increasingly apparent (ACIA, 2005; Nilsson and Döscher, 2013; Wormbs et al., 2017). Following the record Arctic sea-ice minimum of 2007, and even more notably in connection with the 2012 sea-ice minimum, the circumpolar region gained new significance and became a physical and discursive space where global and regional imaginaries both merged and collided. The sea ice had thus moved from being the backdrop to heroic exploration voyages of the 19th century (e.g. Nansen, 1897) to become a heavily mediated resource for actors claiming a stake in the future of the northern polar region.

As a region where few people live or visit for extended periods of time, the Arctic in a global context assumes form and meaning through popular communication channels and most notably through the news media. The section below, 'Media narratives', presents elements of those imaginaries based on a short summary of earlier research, new analyses of major mainstream news outlets and interviews with representatives of news organizations that cover the Arctic region. Today's media landscape is tightly linked to Internet access, and the latter part of the chapter, 'Media cartographies', focuses on the policies and politics relating to the broadband infrastructure that influence the availability and use of online news and communications platforms. The concluding remarks reflect on the potential consequences of the developments observed for the future mediations of the Arctic.

Media narratives

The past decade has featured an increasing body of literature on coverage of the Arctic in various media, often using discourse analysis and frame analysis in their methodological and theoretical approaches (Berzina, 2015; Bushue, 2015; Chater and Landriault, 2016; Christensen, 2013; Davies et al., 2017; Gritsenko, 2016; Hiebert, 2014; Nefidova, 2014; Pincus and Ali, 2016; Reistad, 2016; Steinberg et al., 2014; Stoddart and Smith, 2016; Tjernshaugen and Bang, 2005; Wilson Rowe, 2013; Wilson Rowe and Blakkisrud, 2012). While coverage of the Arctic dates back many decades, the more widespread media (and academic) attention on the Arctic has coincided with the increased attention on climatic change and its consequences for the Arctic. For example, an earlier study based on an analysis of coverage of the Arctic in the Swedish and international news media revealed an increase in the number of articles between the two four-year periods 2003–2006 and 2007–2010 (Christensen, 2013). A surge in media interest after 2007 was visible in a study of Canadian media coverage of the Arctic (Nicol, 2013). On coverage specifically related to climate change and the Arctic, Christensen found that Arctic climate issues were covered in topically diverse ways during this period, and in a way that linked global issues to the Arctic and vice versa. Furthermore, scientific certainty was highlighted in contrast to an earlier common focus of climate reporting on doubt and uncertainty. In hindsight, it is apparent that Arctic climate change had become a 'meta-event' (Christensen, 2013) and the sea-ice minima of 2007 and 2012 served as significant moments with consequences for the media visibility of the Arctic and for linking climate change issues to security frames and geopolitical considerations.

While the number of studies of Arctic media content has grown, studies often focus on a limited number of issues or a national context in order to provide more in-depth analyses. Examples include studies on press coverage of climate issues

(Christensen, 2013; Stoddart and Smith, 2016; Tjernshaugen and Bang, 2005), the conflict–cooperation dichotomy (Wilson Rowe, 2013), the race for resources (Pincus and Ali, 2016), the Arctic Council (Chater and Landriault, 2016; Steinberg et al., 2014) and governance mechanisms (Buurman and Christensen, 2017). A few studies have tried to cast a broad net looking at Arctic coverage more generally, including Nicol (2013) on Canadian media (as mentioned in Chapter 1) and Devyatkin et al. (2017) who used automated identification of topics in documents from the web, mainly in Russian. In adding to these studies, we provide a helicopter view of Arctic coverage in the English-language press in 2007–2015 as well as findings from a frame analysis of two Russian newspapers between 2007 and 2016. This is complemented with insights gathered from in-depth interviews with journalists and editors who work at international and regional news outlets.

English-language newspaper content

To get an overview of how elite, 'quality' or 'broadsheet' newspapers in the United States, Canada and the United Kingdom have covered the Arctic region since the 2007 record sea-ice minimum, a large-scale content analysis of the *New York Times* (USA), *The Guardian* (UK), the *Financial Times* (UK) and the *Globe and Mail* (Canada) was conducted for the period 2007–2015. Every article that mentioned the word 'Arctic' was retrieved from LexisNexis. Using the word cruncher tool in the AtlasTi software, tables were generated of how often different words appeared, and these were used to find the most common meaning-bearing words that might be relevant to how the Arctic region was being discussed. This section presents a selection of the data from the study. In addition, the articles that contained some of the most relevant keywords were reviewed in a qualitative analysis to find the specific context in which these words were used. The analysis begins with an overview of the word count and specific keyword numbers from the *New York Times*, the *Globe and Mail*, *The Guardian* and the *Financial Times*. This is followed by an analysis of word clusters related to a number of specific topics.

Figure 2.1 shows the total number of words published by the *New York Times*, the *Globe and Mail*, *The Guardian* and the *Financial Times* between 2007 and 2015 in articles that mention 'Arctic'. Immediately striking about the graph is that neither the *Globe and Mail* and nor the *Financial Times* saw any significant increase in the number of words published in articles containing the word 'Arctic' in that period. In fact, the *Globe and Mail*, perhaps counterintuitively given the fact that Canada has a vast Arctic territory, saw a gradual reduction in its coverage. The *New York Times*, on the other hand, saw a slight increase in the latter half of the eight-year period. The stand-out newspaper was *The Guardian*, which saw a significant leap in the volume of words published, which began in 2011 but took off in 2014. The total coverage in the *Financial Times* was less than in the other newspapers.

Some interesting patterns emerge when looking at the frequency of use of specific keywords (see Figures 2.2 and 2.3). The major topics that the newspapers have in common are climate and oil. (Oil is discussed in more depth in Chapter 5.)

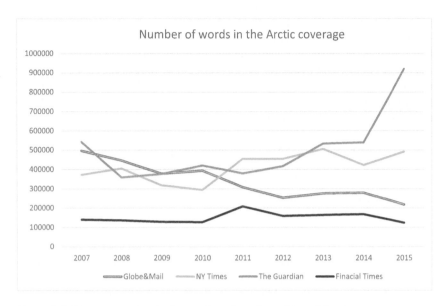

Figure 2.1 Change over time in the number of words used in articles that mention
'Arctic' in the *New York Times*, the *Globe and Mail*, *The Guardian* and the
Financial Times, 2007–2015

Use of the word 'climate' in the articles mentioning the Arctic shows a slight
decline over time in the *Globe and Mail* and a slight increase in the *New York
Times*. In *The Guardian*, the pattern over time is more erratic, with major peaks
and dips. The following are two examples of the use of the word 'climate':

> According to the IPCC, if no action is taken on greenhouse gases, the earth's
> temperature could rise by 4.5C or more. The effects of climate change are
> being felt already, the panel says. The Arctic is warming twice as fast as
> the global average and adverse effects on human activities are documented.
> Impacts of warming have also been observed in other regions and sectors, in
> particular on ecosystems. *The Guardian*
>
> (Aldred, 2007)

> In recent years, climate change has raised the stakes and even non-Arctic
> nations have been looking north for new shipping routes, resource develop-
> ment and a place to express their national ambitions. In response, Mr Harper
> laid out a more muscular approach for Canada in 2007. *Globe and Mail*
>
> (Bird, 2015)

Throughout the eight-year period analysed, *The Guardian* referred to 'climate'
as much as three times more often than the *Globe and Mail* or the *New York
Times*. In addition, just as the total word count for *The Guardian* saw a dramatic

increase in 2014–2015, so too did mentions of 'climate', which more than tripled in that one-year period.

In all three newspapers, 'environment/environmental' was either relatively stable or showed an upswing toward the end of the period (*The Guardian* and the *New York Times*). The frequency of 'oil' varied over time in the *New York Times* and the *Financial Times* and showed an increase in *The Guardian*, especially toward the end of the study period. A closer look at three examples reveals the complex interactions between environmental and economic business interests:

> Greenpeace, which opposes the opening up of Arctic regions to oil exploration, said last night: 'Conditions, whether in April or August, mean drilling is going to be a huge risk for the environment and for investors'. The environmental group disrupted drilling for two days last August when four activists managed to board Cairn's drilling rig, the Stena Don. *The Guardian*
>
> (Webb, 2010)

> The drilling for oil off the west coast of Greenland and the potential environmental hazards were much discussed last summer, but now large-scale mining projects dotted around the rocky coastline are being considered for iron, gold, nickel, platinum and diamonds, to name but a few. With a relatively uneducated and tiny workforce, it is inevitable that Greenland will require thousands of foreign workers to explore and mine these resources—a prospect that concerns many people here in the north because they think their previously 'closed' country, with a population of just over 50000, will be rapidly overwhelmed by people from different cultures. *The Guardian*
>
> (Leonard, 2011)

> The US panel also recommended that the government dramatically increase the Coast Guard presence in the Alaska offshore, describing federal emergency response capabilities as 'very limited'. And it warned the industry will have to make major investments in spill response equipment that is currently lacking in the north. 'Bringing the potentially large oil resources of the Arctic outer continental shelf into production safely will require an especially delicate balancing of economic, human, environmental, and technological factors', the panel concluded. *Globe and Mail*
>
> (McCarthy, 2011)

The sharp increases seen in both 2012 and 2015 in the number of times the words 'ice', 'environment' and 'climate' were used by *The Guardian*, as well as the dramatic increase in the overall level of coverage by *The Guardian* in 2014–2015, are probably linked to two events. The first, in September 2012, was a report that Arctic sea ice had receded to a record minimum level of 18 per cent below the previous record-breaking minimum set in 2007, and 49 per cent below the 1979–2000 average. The second event was the 21st Conference of the Parties to the United Nations Framework Convention on Climate Change (COP21), which

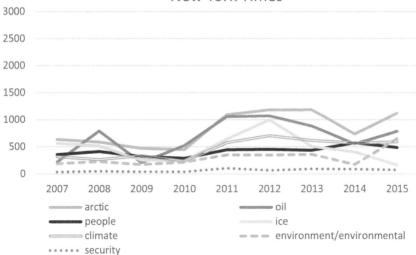

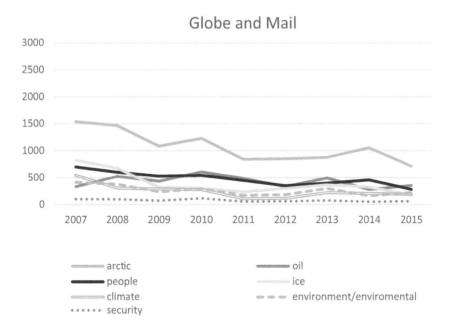

Figure 2.2 Number of times specific keywords were mentioned in the Arctic coverage of the *New York Times* and the *Globe and Mail*, 2007–2015

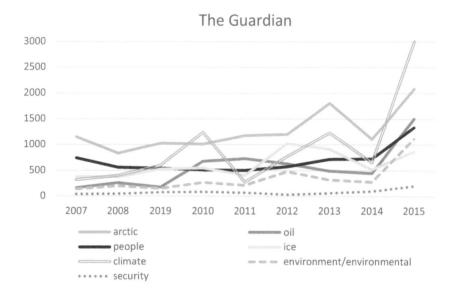

The Guardian

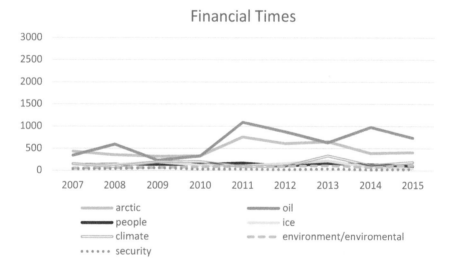

Financial Times

Figure 2.3 Number of times specific keywords were mentioned in the Arctic coverage of
The Guardian and the *Financial Times*, 2007–2015

was held in Paris in December 2015. This led to the 2016 'Paris Accord' on the
global reduction of greenhouse gas emissions in order to slow global warming.
While in *The Guardian* there was a dramatic increase in the amount of attention
paid to the Arctic, it is noteworthy that COP21 did not trigger a corresponding
increase in articles containing the word 'Arctic' in the *Globe and Mail* or the *New
York Times*, both of which are newspapers from countries with Arctic territories.

In addition to the analysis of individual words, an examination of the words used was conducted based on a select number of themes: (a) indigenous populations; (b) protest; (c) technology/media; and (d) profit/business. The results are presented in Table 2.1.

Unsurprisingly, the use of terms to indicate indigenous populations was led by Canada's *Globe and Mail*. The connections between the Arctic and these groups often revolved around threats to traditional ways of life, such as when the *Globe and Mail* wrote:

> Many indigenous peoples depend on hunting polar bears, walruses, seals and caribou as well as herding reindeer and fishing. Access to these species is likely to be seriously impeded by climate warming. Health concerns include an increased accident rate because of environmental changes, such as sea-ice thinning.
>
> (*Globe and Mail*, 2007)

A limited number of articles discussed collaborations between indigenous peoples and scientists:

> In Europe, the Institute of Development Studies is establishing an Indigenous Knowledge and Climate Change Research Network that aims to 'look in greater depth at the learning, exchange and valuing of indigenous knowledge

Table 2.1 Word count for selected themes in the *New York Times*, *Globe and Mail*, *The Guardian* and the *Financial Times*, 2007–2015

	Total	Globe and Mail	NYT	FT	The Guardian
Indigenous populations					
Inuit	3,120	2,358	268	108	386
Indigenous	399	204	123	55	128
Native	1,438	560	542	50	286
Protest					
Protest	1,540	197	396	173	784
Protester/s	401	49	75	55	272
Activist/s	1,613	233	335	169	876
Profit/Business					
Profit	1,660	356	324	477	503
Profitable	244	42	65	70	67
Corporate	952	217	184	201	350
Capitalism	185	35	43	23	84
Corporation	763	179	232	90	262
Technology/Media					
Twitter	250	2	16	0	232
Facebook	458	85	107	73	193
Blogs	1,750	0	1744	1	6
Radio	1,008	282	313	49	364
Television	1,316	53	30	1	224

on climate change', according to Blane Harvey, one of the project's three directors. North American scientists are collaborating with Inuit leaders to gain an understanding of changes in the Arctic and the potential impacts on native livelihoods. *New York Times*

(Walsh, 2011)

In general, however, references to indigenous peoples were few, which is in line with other studies of how the Canadian press covers indigenous issues (e.g. Stoddart and Smith, 2016). The use of words in relation to the second theme, 'Protest', saw a similar domination by one outlet, this time *The Guardian*. The newspaper was the leading user of all three protest-related words that we searched for, with all the others far behind.

The Guardian was also the primary outlet for words in relation to profit/ business. A more detailed look at the themes of the coverage, however, showed these mentions to be largely rhetorical afterthoughts. Akin to the issue of native or indigenous populations and protest, addressing questions of corporations, profits and capitalism in relation to the Arctic appears not to have been a priority, despite the fact that a significant number of issues that link the Arctic to environmental issues and climate change have a direct connection with private sector businesses and profit-led industries such as petroleum. While the word 'oil' in all newspapers combined was mentioned 20,822 times, adding its variants 'gas' and 'petroleum' nearly doubles this rate and reaches around 38,000. Words corporations, capitalism and profit, on the other hand, garnered slightly over 3,800 mentions.

Finally, given the discussions in recent years about the role of the Internet and social media in influencing the national and international news flows, it was interesting to examine the extent to which various media platforms and systems were mentioned in articles where the Arctic was named. What we see in these numbers is that other forms of media—but social media in particular—were not factors in issues of substance in stories from the Arctic. (The high number of mentions in blogs by the *New York Times* is probably linked to its practice of using a blog format in its reporting, leading to the word being used in many articles, but not in the body of the text.) Twitter and Facebook, which are often discussed as important sources of news and information, were largely ignored, as was television.

Russian newspaper content

To map the Russian media discourse, 822 articles in *Rossiyskaya Gazeta* and 127 articles in *Novaya Gazeta* were analysed, focusing on the major frame in each article. The articles were retrieved from the database *Medialogia* using the search term 'Arctic' (in various variations in the Russian language including noun and adjectival forms) for the years 2007–2016. Each article was coded using a set of overall frames from Buurman and Christensen (2017) adapted from studies on how media frame scientific issues (Nisbet and Scheufele, 2009). Details of the study are presented in Klimenko et al. (2019). An overview of the results is provided in Figure 2.4. Examples of the topics from the media coverage relating to the different frames are presented in Table 2.2.

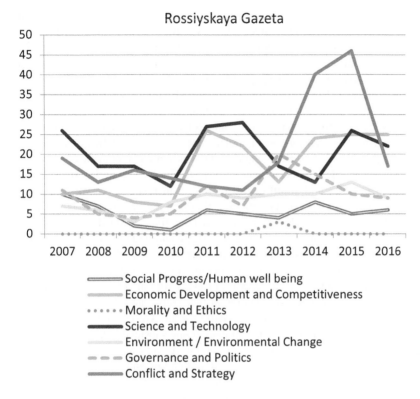

Social Progress/Human well being
Economic Development and Competitiveness
Morality and Ethics
Science and Technology
Environment / Environmental Change
Governance and Politics
Conflict and Strategy

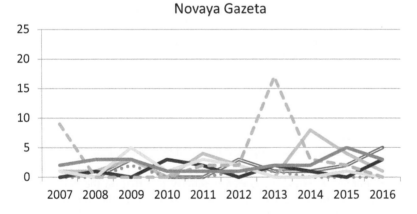

Figure 2.4 Frequency of different framings in *Rossiyskaya Gazeta* and *Novaya Gazeta*, 2007–2016

Rossiyskaya Gazeta is the daily government newspaper of record, which mirrors the official Russian voice, while *Novaya Gazeta* is an independent journalist-owned newspaper published three times a week and famous for its investigative journalism. The amount of attention paid to the Arctic differed substantially between the newspapers, in terms of both volume and how the region was framed. Recurring frames in *Rossiyskaya Gazeta* were conflict and strategy, science and technology, and economic development. While science and technology received rather neutral and consistent coverage throughout the ten-year period, focused on various polar expeditions, scientific achievements and developments, as well as the state of the Russian icebreaker fleet, the other two frames changed in focus and tone over time. The conflict and strategy frame was visible in 2007 and significantly increased in response to western reaction to the planting of a Russian titanium flag on the seabed at the North Pole in 2007. In 2007 and 2008, a number of articles focused on the potential for conflict over territory and resources in the Arctic and underlined growing competition between the Arctic states. For instance, Nadezhda Sorokina wrote that the 'division of the Arctic will be the beginning of a new re-division of the world' (Sorokina, 2007), while Darya Yurieva wrote about US and Canadian 'harsh statements about Russian claims in the Arctic' and the participation of the 'mighty among the nations' in the 'battle for the Arctic' (Yurieva, 2007).

The dominance of the conflict and strategy frame subsided in later years, as the focus instead shifted to cooperation between the Arctic states, diplomatic rather than military resolution of competing territorial claims and governance, including UNCLOS and the Arctic Council. The language also changed to a greater emphasis on 'absence of disputes in the Arctic' (Feschenko, 2011), 'international dialogue' in the Arctic and opportunities for bilateral cooperation between Russia and various states (Smolyakova, 2011). In 2014–2015, however, following the annexation of Crimea and subsequent western sanctions, the conflict and strategy frame was resurgent. Although the theme of potential conflict in the Arctic did not return to the discussion, coverage started to focus on military developments in the region, such as various military exercises, the construction of military assets in the Arctic and the views of military personnel on developments in the Arctic region. By 2016, however, the extreme level of attention on military issues had subsided once again.

Attention on economic development mirrored both market changes and geopolitics. For example, in years of high hope for the development of Russian offshore resources, *Rossiyskaya Gazeta* regularly reported on the potential for the development of oil and gas resources on the Arctic shelf, the strategies of the state companies, Rosneft and Gazprom, for Arctic resource development and cooperation between state companies and international players in the Russian Arctic. The discussion post-2014 focused primarily on the impacts of western sanctions on Russia.

Compared to the western English-language media, *Rossiyskaya Gazeta* paid only limited attention to issues related to climate change and the environment, and the notion that climate change is human-induced was still being questioned by some experts in Russia. Increased economic opportunities linked to offshore resource extraction and shipping were emphasized in discussions on observed changes in the environment, such as receding sea ice, as illustrated in the following passage from *Rossiyskaya Gazeta*: 'This [melting of the ice] is good for the Russian economy because it ensures the full return of Russia to the Arctic' (Krivoshapko, 2016). Another focus of the discussion within the frame of the environment/environmental change was 'cleaning up' the Arctic islands after significant levels of pollution during the Soviet era.

The reporting in *Novaya Gazeta* focused less on conflict and strategy than *Rossiyskaya Gazeta*. In relative terms, it also had more articles with the frame social progress and human well-being, especially toward the end of the study period. In this frame *Novaya Gazeta* often raised questions about difficult living conditions in the northern communities, which were omitted from *Rossiyskaya Gazeta*'s coverage. *Novaya Gazeta* also paid more attention to the environment and environmental change, with a focus on potential ecological damage linked to the development of oil and gas, and mining industries, rather than on the economic opportunities presented by climate change. Also notable is the number of articles with a governance and politics frame in 2013 when *Novaya Gazeta* ran significant coverage of the *Arctic Sunrise* case, in contrast to *Rossiyskaya Gazeta* which published only four articles about it.

More in-depth and longitudinal studies would be needed in order to pinpoint what factors have been driving the Russian media's coverage of the Arctic and how it compares to the western international media reportage. A large-scale study by Devyatkin et al. (2017) provides some additional context. Using automated identification of topics from the web they identified the most prevalent topic as hydrocarbon production in autonomous areas and offshore in the Arctic Ocean, followed by: transport issues (including Northern Sea Route); public steps to support northern regions and indigenous peoples; scientific expeditions to assess climate change; government support programs and infrastructure development; tourism in the European north; marine scientific expeditions; protection of species; deployment of the Arctic forces group; and politics of offshore area ownership. Furthermore, Devyatkin et al. note that 'international media pay more attention to conflict issues rather than the development of infrastructure, tourism or environment protection' (p. 335). This is in line with an observation by Sergunin and Konyshev (2016) that the drivers of the current discourse in Russia are competition for resources and the control of the Northern Sea Route. Yet, contrasting Cold War geopolitics with today's economic and strategic interests could be falsely dichotomous considering the complexities inherent in today's geopolitical dynamics. Within Russia, rifts remain between those who champion internationalist and cooperative discourses and those who favour expansionist and imperialist ones (ibid.). At international forums, Russian representatives emphasize cooperation and the supremacy of international law. Tensions between cooperative-pragmatist views and more 'imperialist' assertions are also observable in the positionalities of other international actors with a stake in the Arctic, all depending on the political conjunctures that characterize the regional and international agendas and how they

Table 2.2 Examples of issues discussed within different frames in the Russian newspapers *Rossiyskaya Gazeta* and *Novaya Gazeta*, 2007–2016

Social progress/human well-being	Economic development and competitiveness	Morality and ethics	Science and technology
Cultural events	Oil and gas sector developments	Indigenous population in the Arctic	Icebreakers
Socio-economic strategies of various regions of the Russian Arctic	Arctic resources	RAIPON	Search and rescue centres
Housing problems in the Russian Arctic	Resources of the Arctic shelf		Scientific and technological developments
Healthcare system in the Russian Arctic regions	Arctic tourism		Expeditions to the North Pole
Problems of infrastructure in the region and its influence on the well-being of the population	Cooperation/agreements between various oil development companies		Infrastructure projects/port infrastructure
	Shipping		Development of Arctic aviation
	Development of the Northern Sea Route		Construction/development of equipment for the oil/gas industry
	Fisheries		Scientific and research cooperation
	Economic sanctions		

Environment / Environmental Change	Governance and Politics	Conflict and Strategy
Melting of Arctic sea ice	Cooperation in the Arctic Council, ministerial meetings	Potential for conflict in the Arctic region
Environmental damage/pollution	BEAC and other institutions, meetings at various levels	Territorial claims (when mentions possible conflict)
Ecosystems	UNCLOS and territorial claims (where no potential for conflict is mentioned)	Military developments in the Arctic region, including exercises, presence of military forces, fleet, etc.
Weather and temperature changes in the Arctic	Bilateral visits and bilateral cooperation initiatives	Development of border guard facilities in the Russian Arctic
Climate change	Various state strategies and laws to regulate the Arctic	
Global warming	Legal disputes (*Arctic Sunrise*)	
'Spring cleaning' of the Arctic area/Russian Arctic islands		
Permafrost melting		
Nature resorts in the Arctic		

are affected by shifting geographic factors and their mediation. We return to these questions in Chapters 4 and 6.

Perspectives from the field

To put the analysis of western media content in context, open-ended interviews were conducted with journalists at two major international newspapers: *The Guardian* and the *Financial Times*. Our empirical material also includes perspectives presented at the Arctic Circle Assembly panels organized by ourselves and others in 2016 and 2018. The insights that journalists in the field shared illustrate some of the material and political/professional dynamics that affect decision making and the media logic that is in play behind the stories that reach readers.

The circumpolar region has made the headlines in the international news media primarily due to its melting sea ice, decorated with dramatic pictures of age-old glaciers perishing chunk by chunk and starving polar bears. Fiona Harvey, who has been the environment correspondent at *The Guardian* since 2011, noted that climate change has made the Arctic an international news story:

> Arctic science is absolutely fundamental to climate change in every way and from a journalist's perspective, the Arctic is something people really connect to even if they have never been there, partly because the ice caps became beacons, and also because of the images of polar bears and all that.
>
> (Harvey, 2018)

From a readership and audience perspective, it is significant that the Antarctic was linked with another widely mediated environmental issue, the hole in the ozone layer. Harvey noted that this phenomenon had already conditioned readers globally to perceive polar science as having planetary significance:

> The Arctic became an imaginative vehicle. . . . The Arctic and the poles speak to the imagination in unique ways. It was not that long ago when we had the first expeditions to the poles. These are still very alive in people's imagination. It's exotic and almost unimaginable; and its unimaginability is its great strength. A vast white desert. That blankness provides a canvas onto which people's own imaginings get inscribed. . . . Talking about the Arctic became a shorthand for talking about climate change.
>
> (Harvey, 2018)

The record sea-ice minima of 2007 and 2012 and the changes in the Arctic sea ice provided 'set pieces' and new news hooks for journalists to write about the Arctic in the mainstream media.

Our analysis of the news content as well as the interviews indicate that journalists and editors judge that international readers and audiences are less interested in social or regional and local issues, unless these are weather events or environmental disasters, and more interested in the scientific dimension or in the global implications for finance and trade, such as shipping routes and oil drilling. Fiona Harvey remarked that 'people are bored with seeing pictures of politicians, mostly men in suits, so when picking pictures for the Arctic or other environmental issues, we keep that in mind. People love looking at pictures of wildlife'. She added:

> Arctic has a mysticism; something greater than people's imagination compared to other environmental realms; it has an enormous emotional pull. It's remote, very few people visit it. It brings together the emotional and the imaginative. . . . and we should use that as journalists. . . . Some animals like polar bears are charismatic megaphones for environmental issues. Journalists are often shy about using that emotional connection, but we shouldn't be because what is at stake is not just a dry scientific subject or a political subject. What we are talking about is the future of the planet and it is an emotional subject.
>
> (Harvey, 2018)

Pilita Clark, Associate Editor and columnist at the *Financial Times*, who worked as its environment correspondent between 2011 and mid-2017, offered similar observations, noting that climate change is the main issue that has put the Arctic on the news agenda: 'The Arctic is the canary in the coalmine, it is where we see the effects of global change most potently and obviously' (Clark, 2018). While the Arctic has become of more interest to non-specialist journalists, because of the multitude of trade and geopolitical implications, Clark said that the fact that it is difficult and expensive to get there explained why it remains under covered in the international news media. As Google and Facebook increasingly swallow up advertising and the newspaper business becomes more fragile, the environmental desks at many newspapers shrink or disappear altogether. However, Clark noted that climate change is becoming a subject of increasing interest, especially to younger readers and audiences—and younger audiences are the ones that media organizations are trying to hang on to. In that regard, using climate change as a news hook to cover Arctic issues, including the social, political and economic dimensions as well as questions other than climate change itself, remains a strategic choice. She mentioned new scientific information or a new record low in sea-ice measurements as important drivers of the news coverage. One aspect that can draw people's attention is the relationship between the melting ice and weather patterns.

While generally similar to other international newspapers, Clark mentioned that the *Financial Times* is more focused on business than other papers, such as issues related to shipping and who is likely to benefit from new routes: 'More granular stories in terms of what is happening on the ground, if people are getting excited or worried as the case may be from a business and financial perspective' (Clark, 2018). She also highlighted that the perspective was mainly global, noting that the Arctic 'has been covered from a global perspective, in terms of the consequences of its change on trade and on people elsewhere' and added that 'local aspects have not been covered that much partly because it is difficult to get there'. Regarding the changes in the news coverage in recent years, Clark mentioned changing attitudes to climate change sceptics since 2011:

> Climate [change] sceptics and deniers have been downgraded. I was never asked to include this and that for balance in my stories or deniers for balance. But if say there was a report issued by a conservative think tank, then the editor would perhaps say ok let's have a look at it, see what it is all about. I think also the Paris Accord made a difference.
>
> (Clark, 2018)

Clark also mentioned that the BBC has recently issued new guidelines to the effect that a climate change denier or sceptic is no longer required to appear on screen for balance in any report on climate change. This followed complaints to the BBC after radio journalists failed to counter comments that the Earth was not warming. The pressure accumulated, according to Clark, adding that all the extreme weather in the past summer also played a part.

Clark also commented on the scalar dimension of media coverage:

> It has been covered from a global perspective, in terms of the consequences of its change on trade and on people elsewhere. People are interested in the scientific elements and science stories, in the ways in which the Arctic is linked to climate change, to trade. But local aspects have not been covered that much partly because it is difficult to get there.
>
> (Clark, 2018)

She mentioned that the lack of in-depth focus on the Arctic or on climate change in the mainstream newspapers is compensated for by online specialist news organizations such as *Climate Home, Carbon Brief* and *Artic Today*. However, they are only able to do this if some sort of philanthropic or charitable funding or some other form of funding is available: 'So it's not like they have a very stable business model in place' (Clark, 2018).

For Arctic news sites, business models are evolving to include institutional subscriptions from educational and research institutions as well as building alliances among small online news outlets, as is discussed further below based on interviews. While the future of news production, following declining revenues for some big media and the increasing prominence of smaller and regional media, remains an open-ended question, current developments on the regional news scene in particular seem promising. Nonetheless, bringing the Arctic into the international news media limelight beyond polar bears and melting ice remains a challenge. As Leanne Clare, senior manager of communications at WWF's (World Wide Fund for Nature) Arctic Programme suggests, 'When the symbol gets bigger than the region itself and people don't realize that the polar bear is just one piece of a whole diverse web of life in the Arctic, then it can become almost a barrier' (cited in Breum, 2018).

Regional close-ups

The appearance in the past decade of media that specialize in news about the Arctic has been a major shift in the media landscape. One prominent example is *Arctic Today*, which describes itself as 'an independent news source in partnership with media organizations from around the circumpolar North' (*Arctic Today*, n.d.). Another news outlet is the *Barents Observer*, which publishes in English and Russian. It was initially a project run by the Norwegian Barents Secretariat and sponsored by the Norwegian government, but is now an independent journalist-owned platform (see below). Another news platform with ambitions of circumpolar reach is *Eye on the Arctic*, which is run by the Canadian National Broadcasting Corporation, and publishes its website in English and French. A fourth outlet, *High North News*, describes itself as 'an independent newspaper published by the High North Center at the Nord university'.

The emergence of these news outlets is linked to increasing interest in the Arctic region but also to the technical capacity to publish online, which has improved in the past decade. Moreover, their reporting builds on networks of collaboration among journalists covering the Arctic from various local and national perspectives, through which stories can be shared. These outlets can therefore also serve as platforms for gathering similar experiences from different local contexts. The regional news outlets have some overarching challenges in common linked to economics and politics. A close-up examination of two prominent examples, *Arctic Today* and the *Barents Observer*, illustrates some of these dynamics.

Arctic Now/Arctic Today

Arctic Today was launched at the end of January 2018 as a follow-on to another online outlet, *Arctic Now*, which began publishing in October 2016 as a specialized online news site affiliated with the Anchorage-based *Alaska Dispatch News*. *Alaska Dispatch News*, in turn, was the product of a merger between the *Anchorage Daily News*, a traditional newspaper published in Anchorage, and the online news blog *Alaska Dispatch*, which the *Colombia Journalism Review* described as a 'regional reporting powerhouse . . . that fights aggressively for online ad dollars' (Canyon Meyer, 2010). In 2014, the *Anchorage Daily News* was bought by *Alaska Dispatch* and the merger of the two sites led to the launch of the consolidated outlet *Alaska Dispatch News* (Alexander and Puh, 2014; Medred, 2014). While the merger raised questions about the role of traditional print publishing in an era of online news, as well as questions about economic viability (Devonne, 2014), *Alaska Dispatch News* quickly gained in popularity and gave Alaskan news a presence through its website, adn.com.

Our interviews with Alice Rogoff, the owner of *Arctic Today*, Kristia DeGeorges, the editor-in-chief of *Arctic Today*, and Kevin McGwin, a Denmark-based correspondent of *Arctic Today*, yielded further insights. Arctic Now was launched as an online circumpolar news source with ambitious outreach goals at the Arctic Circle Conference in October 2016. It gathered its news from *Alaska Dispatch News*, along with material from wire services and international newspapers such as Reuters, Tribune Content Agency, the *New York Times* and the *Washington Post*, as well as from five partners around the Arctic—the *Independent Barents Observer* and *High North News*, both from Norway; *The Arctic Journal*, based in Greenland; and the *Iceland Monitor* and *Nunatsiaq News*, based in Iqaluit, Canada. According to Alice Rogoff, the model was meant to gather Arctic news from all these sources and run a subscription-based, comprehensive website, which was online until January 2018. *Arctic Now* also occasionally published material written exclusively for the site and solicited opinion pieces (DeGeorges, 2017, 2018).

The business venture witnessed a major disruption in August 2017 when *Alaska Dispatch News* filed for bankruptcy. Alice Rogoff told *Anchorage Daily News* in 2014:

> We had worked hard to help illuminate the issues of the day and provide a platform for views from across Alaska . . . yet like newspapers everywhere, the struggle to make ends meet financially eventually caught up with us. I simply ran out of my ability to subsidize this great news product. Financial realities can't be wished away.
>
> (Zak, 2017)

Reverting to its old name, *Anchorage Daily News* continued publishing under new ownership, but Rogoff took the core staff from *Arctic Now* with her when she launched *Arctic Today* in January 2018. In terms of content, the focus was similar—news of potential interest to a circumpolar audience with a stake in the Arctic. Also like *Arctic Now*, the content is provided behind a paywall. In her inaugural email following the launch of the site, on 29 January 2018, Rogoff asked readers to become paying subscribers: 'As with so many news sites, the only way we can continue as a business is to ask you, the reader, to pay us for what we do!' *Arctic Today* works with the same constellation of circumpolar partners as its predecessor, apart from *The Arctic Journal* which ceased publication in June 2017. As of December 2018, *Arctic Today* publishes email newsletters five days a week and maintains a website with news, feature articles and commentary aimed at an audience with circumpolar interests. Rogoff noted that all the user data and statistics indicate that its readers are serious news consumers, mostly from outside the region, with an interest in the Arctic, and that their base is growing. The aim is not only to expand their viewership but also create it by catering to a diverse variety of sectors such as academia, business and finance, politicians and the general public (DeGeorges, 2017, 2018). One type of journalism that is lagging behind is cultural journalism and community news, while scientific news and geopolitical accounts such as opening airports in Greenland or Chinese investment in the Arctic dominate the agenda. *Arctic Today* recently added a travel section to appeal to those outside the region 'who are interested in the ice, polar bears and similar features'. Rogoff mentioned that *Arctic Today* has more journalists on site covering the region and producing original stories and is engaged in a new initiative to form a cooperative alliance with small news outlets, blogs and publications to eventually include NGOs and their newsletters: 'We are trying in every way possible to organize amongst ourselves to form a one-stop-shop for readers because otherwise it is like a needle in a haystack to find readers' (Rogoff, 2018).

Rogoff indicated that while difficulties persist, the growing number of institutional subscribers and of alliances between similar-sized news outlets with similar scope helps the business model to sustain itself. However, she emphasized the need for more public and institutional support, and for grants to support the networking of such outlets to form more robust constellations as one particularly desirable mechanism. At the Arctic Circle 2018 panel, she commented: 'The line between for profit and non-profit journalism has frankly disappeared because for profit journalism doesn't make a profit anymore and non-profit journalism means dependence on someone else's journalism', adding, during our interview that 'the world is less dependent on the journalism of international legacy media. It is our job now to provide Arctic news' (Rogoff, 2018).

Commenting on the potential for political pressure to be put on journalists covering the Arctic, Rogoff highlighted how contentions around climate change in the United States have had an effect on their way of reporting:

> In the media market in Alaska, there are a large number of people who will turn off when they see a story about climate change because they don't believe in climate change. So you have to remember who you are writing for. That's a depressing fact.

She also lamented how a lack of resources meant that reporting mainly focused on news, making it difficult to make up for a collective absence of voice for people living in the north: 'We struggle with so few resources that we cover just the bare bones. What you don't get is enough resources for features that capture the imagination of people'.

The Barents Observer

The *Barents Observer* describes itself as a 'journalist-owned online newspaper covering the Barents Region and the Arctic'. According to its website, it has 'a devotion for cross-border journalism, dialogue and mutual understanding'. The story behind the current website, which was launched in the autumn of 2015, illustrates the political challenges involved in Arctic reporting. Our first interview with its editor-in-chief, Thomas Nilsen, took place in October 2015 in a small office into which he and his colleague, Atle Staalesen, had just moved to set up a new online news outlet: the *Independent Barents Observer*. One of the walls of the office was decorated with a poster of the Norwegian declaration on the rights and duties of the editor and Act relating to editorial freedom in media, proclaiming that 'the editor shall lead the editorial operation and take decisions on editorial questions', and that '[t]he owner of the media enterprise or the person who leads the enterprise on the owner's behalf may not instruct or overrule the editor in editorial questions' (ACT-2008-06-13-41, see Norsk Redaktørforening, n.d.). In early 2015, Thomas Nilsen had been dismissed from his position as editor of the *Barents Observer*, which was then part of the Norwegian Barents Secretariat and owned by the three northernmost counties of Norway: Finnmark, Troms and Nordland. The reason given was that Nilsen had published a letter of protest signed by all four staff members at the *Barents Observer* about its owners meddling in the outlet's content.

As Nilsen detailed during our interviews, this was the end result of escalating tensions that began in March 2014, when the Russian Consulate General in Kirkenes launched a major public outburst against the *Barents Observer*, calling it anti-Russian and criticizing it for writing about the annexation of Crimea. Russian-Norwegian relations became tense in a situation where friendly High North cross-border relations were important to maintain, and the Consulate General saw the *Barents Observer* as a megaphone for regional politicians. According to Nilsen (2018a): 'Things reached boiling point when the local politicians told us to be careful not to tarnish Barents values'. Nilsen and his colleagues responded to this challenge by publishing a letter of protest on 21 May 2015, following which Nilsen was dismissed on the spot for alleged disloyalty.

The situation made waves in Norwegian media circles and within the Norwegian Ministry of Foreign Affairs, and the Norwegian Barents Secretariat was forced to rehire Nilsen. After reaching a settlement, Nilsen and the entire editorial staff left and the Secretariat eventually decided to end publication of the *Barents Observer*. It is now only available in the Norwegian national archive (*Barents Observer*, 2018). News about the closure of the paper and Nilsen's dismissal reached the international media when the Norwegian Broadcasting Corporation, based on an anonymous high-level source, claimed that the Russian security services had demanded Nilsen's removal (Strand, 2015).

However, Nilsen and Staalesen continued their reporting, initially under the banner of the *Independent Barents Observer*, which then became the *Barents Observer* once more, still with the aim of facilitating the exchange of information across the border by making news about Russia available in northern Norway and vice versa. Thomas Nilsen stated that 20 per cent of the outlet's readers are from Norway, 20 per cent are from among indigenous populations, 20 per cent are from Russia and the rest are from Scandinavia and Asia, with a growing number of readers from East Asia (Nilsen, 2018b). Nilsen remarked that the future of journalism is more local, with more contextualized coverage involving cooperation across small and medium-sized outlets. Getting reliable information from the Russian Arctic is still a problem and the entry of foreign journalists into the country is restricted through the visa process. In addition, some, like Nilsen himself, are banned from entering the country altogether.

In contrast to growing political challenges, new technologies have made reporting easier. In the Arctic Circle panel discussion, Nilsen reflected on how new technologies have changed the conditions for Arctic journalism:

When I started in Murmansk in the late days of the Soviet Union, there was one way to communicate with my editorial desk at home in Norway and that was to order a telephone. That had to be done 24 hours in advance. Sometimes the phone line lasted for five minutes, other times for 30 seconds, and then it was to order another phone line and wait for 24 hours. Nowadays, I have the ability to contact people on the tundra in the Nenets region far north of the Arctic Circle by simple, easy [mobile phone technology]. That is key to Arctic journalism. You don't need a lot of money to run a newspaper today. We have close to zero, but we have an international audience and that is due to the fact that I can travel around with my nearly four-year old half-broken Samsung but with a good camera. I do interviews with this and take photos. So, we are able to be in the Arctic and spread the message to the rest of the world, with very little funding. The only thing we need is a good contact network.

(Nilsen, 2018a)

Political challenges remain, however. For example, Nilsen cannot pay Russian journalists for stories because they would risk being branded 'foreign agents' by the Russian authorities. For him the politics of journalism is geopolitics, steered by Washington, Moscow, Oslo and so on. Being on the stop list of people who cannot enter Russia, and naturally wanting that status changed, he relayed an exchange that illustrates the situation:

The Russian Consulate tells me that if you write several articles that tell what Russia is all about—good journalism is telling the story of 1000 years of peaceful relations between Russia and Norway—if I do that, then they might reconsider. That is direct interference in freedom of the media in the High North.

(Nilsen, 2018a)

Media cartographies

Today's media landscape is tightly connected to the growth of the Internet and mobile communications. While much of the Arctic has remained off-grid for longer than more populated regions, Internet access is now growing. This also affects the kinds of stories that can reach either across or beyond the region. Eilis Quin, editor of the website *Eye on the Arctic* at the Canadian Broadcasting Corporation, illustrated the point at a panel discussion on Arctic journalism at the 2018 Arctic Circle Assembly:

In North America, there has always been extremely strong regional coverage, in Alaska with the Alaska Public Radio Network and in Canada the CBC North, and strong local news organizations covering the villages and what is going on at the local level. What technology has let us do within the last 10 years is really focus on pan-Arctic coverage.

(Quinn, 2018)

Eye on the Arctic went online in 2010. It is based on partnerships with northern-based media across the Arctic and the sharing of stories on a non-profit basis. According to Quinn: 'The interest from all these organizations has been to share what is going on in our different Arctic regions and that is one of the things that technology has enabled'. She also highlighted how social media has made it possible for people living in the Arctic to respond to comments and views from the outside, providing media access to previously marginalized voices: 'If we go back to the EU seal ban and the impact that had on Inuit communities in Canada, it was catastrophic, and the legacy of that continues today. . . . There wasn't Arctic media that could amplify those voices'. She contrasted this with how the reaction to a video clip of a starving polar bear that had been linked to climate change went viral and really did make a difference:

> Twitter erupted with Inuit voices across the Arctic saying "what they are saying is not true. This bear is sick and old, not [dying] because of climate change". That then got into the mainstream media because there was this complete flood on Twitter.

She said it was really interesting to see how agendas from non-Arctic regions that often get into the mainstream media no longer go unchallenged:

> Social media is turning all that around and nobody can say that they don't know. These are some of the voices that are amplified. It is still not enough but it's really important to see this going on more and more.

Future visions and imaginaries of the Arctic may thus depend as much on the development of communications infrastructure in the north as on news and cultural industry production elsewhere. The current state of Arctic Internet infrastructure is summarized below.

'Internyet': from disconnection to connection in the Arctic region

According to local legend, when residents in the northern regions of Russia are asked about their access to the Internet, their response is an ironic, resigned 'Internyet'. For these Russian citizens, poor access, low speeds and high cost are a fact of life. That residents of the Arctic region—in Russia, Iceland, the United States, Canada, Greenland, Sweden, Norway and Finland—have experienced more sporadic, lower-quality access to the Internet than more southerly residents in those countries is no secret. In addition, that residents can technically have 'access' to the Internet—through fixed broadband or mobile broadband—masks the fact that there can be significant differences in the quality and cost of such access, with huge variations in download and upload speeds as a prime example.

Based on reports from the Arctic Economic Council (AEC) (2016) and the Arctic Council (2017), the state of play for the Arctic region regarding the challenges and implications of providing high-quality, high-speed, low-cost broadband access to the region are set out below, including a brief, country-by-country summary of current or planned policies to increase access to broadband.

Arctic broadband as a policy goal

At the 2015 Ministerial Meeting of the Arctic Council in Iqaluit, Canada, the ministers of the Arctic states agreed to form a Task Force on Telecommunications Infrastructure in the Arctic (TFTIA). The rationale for the decision was outlined in the Iqaluit Declaration (Arctic Council, 2015), which noted 'the importance of telecommunications to Arctic communities, [as well as] science, navigation and emergency response', and set a goal to 'develop a circumpolar infrastructure assessment as a first step in exploring ways to improve telecommunications in the Arctic'. In short, telecommunications, including broadband access, was defined as a key factor 'for sustainable development in the Arctic' (Arctic Council, 2017). Over a two-year period, the TFTIA, 'worked to assemble and assess information about the available telecommunications infrastructure in the Arctic and the present-day needs of users living, working, or traveling in the Arctic', to review the technologies available to meet the needs of these users and to identify the 'gaps in the infrastructure . . . essential in providing acceptable connectivity' (ibid.).

As a result of the formation of the TFTIA, *Telecommunications Infrastructure in The Arctic: A Circumpolar Assessment* was published by the Arctic Council in 2017. In the introduction, the TFTIA notes the 'enormous variations in the population densities and associated telecommunications infrastructure and services present across the Arctic'. While certain Arctic countries and regions, such as Finland, Iceland, Norway, north-western Russia and Sweden, are more densely populated, and thus have greater access to a variety of telecommunications services, 'the vast expanses of the Canadian, Greenlandic, Russian, and US Arctic have extremely low population densities often with lesser availability'. Specifically, the report identified the 'needs of indigenous peoples and local communities' and the fact that 'improved connectivity in the Arctic supports better access to education, healthcare, and commerce, as well as enhancing citizens' participation in civic life and improving delivery of services'. In particular, telecommunications was regarded as of particular importance to indigenous peoples, 'in maintaining and preserving their cultures and livelihoods' (Arctic Council, 2017).

The TFTIA called high-speed Internet/broadband a 'transformative technology' that 'enables a range of life-enhancing technologies and facilitates convenient and cost-effective communication among family and friends'. Of specific relevance to a discussion of how broadband access can affect the ways in which environmental questions are addressed, and how technology can help to connect, for example, activists in disparate locations across the Arctic region, the study noted how high-quality broadband 'also helps to break down the barriers of distance and time, potentially allowing Arctic residents to more actively participate in economic and civic life far beyond their geographic locations'.

Similar conclusions on the transformative potential of broadband were reached in *Arctic Broadband: Recommendations for an Interconnected Arctic*

(Arctic Economic Council, 2016). In its second report on the topic, the AEC drew attention to the fact that, despite the obvious social, economic and cultural benefits of broadband access, 'broadband deployment and adoption across the globe have not been uniform', and that 'one region in danger of being left behind is the Arctic'. Interestingly, and perhaps somewhat ironically, the AEC—which is primarily a business-oriented organization—was more overt in outlining the relationship between broadband access and political participation than the inter-governmental Arctic Council. In its report, the AEC wrote that:

> Broadband helps to prevent political isolation and to enable political engage-ment and participation. Broadband improves access to information and allows individuals to take part in the political process regardless of one's physical location. Broadband also improves community involvement by allowing stakeholders to exchange ideas, meet, and collaborate in order to work through issues confronting the community.
>
> (Arctic Economic Council, 2016)

While largely advocating private sector governance of the provision of broadband services, the AEC argued that, 'if public-private partnerships are not practica-ble, governments should use public resources to finance and operate broadband facilities to offer services on a carrier-neutral and cost-based wholesale basis to help spur adoption'. In order to spur adoption in communities that were getting broadband for the first time, the AEC recommended that, 'private sector players, NGOs and governments at all levels should be encouraged to work together to promote digital literacy programs'. In addition, these actors should, according to the proposal, 'coordinate with schools, libraries, community centers and other anchor institutions to leverage their presence to spur digital literacy and broad-band adoption'. While the AEC position was clearly one intended to promote broadband adoption in order to benefit the private telecommunications sector, recommendations on media literacy are a valuable reminder that simple access is not sufficient to guarantee rapid adoption and use in communities previously cut off from high-speed services. This, in turn, is a reminder to contextualize techno-determinist discourses in which information and communications precariousness are immediately overcome with the introduction of new services—a perspective that ignores the social, economic, political and cultural factors that can influence such use. Even with this in mind, however, there can be no uptake without access, and the following section assesses the current 'state of play' regarding broadband access in the Arctic regions on a country-by-country basis.

The state of play: broadband in the Arctic region

There is no up-to-date consolidated summary of Arctic Internet access. The following brief summary is compiled based on a number of sources (Arctic Economic Council, 2016; Government Offices of Sweden, 2017; TeleGeography, 2017; Broadband Now, 2018; Nilsen, 2018c; Northwestel, 2018; Staalesen, 2018; Tusagassiuutit, 2018).

Canada

In December 2016, the Canadian Radio-television and Telecommunications Commission (CRTC) announced a new regulatory framework in which fixed and mobile broadband Internet access were classified as a basic telecommunications service. The original targets were to provide access to 50 Mbps download and 10 Mbps upload to households and businesses and along major roads. A proportion of the costs would be covered by a tax on telecommunications revenues, leading to the creation of a fund that would provide up to CAD 750 million. In addition, starting in 2017–2018, Investing in Canada earmarked CAD 2 billion in funding for rural areas for investment in infrastructure, including improved Internet connectivity. One concrete example of these developments was that in October 2018, residents of Iqaluit in Nunavut had their Internet access speeds tripled when Northwestel announced that the city would get access to its high-speed satellite network. Known as 'Tamarmik Nunaliit' (Inuktitut for 'every community'), the new network has a capacity 20 times greater than the previous satellite. It is worth noting, however, that the cost of access is unaffordable: a 15 Mbps package costs CAD 129 per month. The Canadian government invested CAD 50 million in the Northwestel project to improve coverage in Nunavut. It is anticipated that the faster service will be available to all 25 Nunavut communities by the end of 2019.

Finland

In July 2010, Finland became the first country in the world to declare broadband access to be a legal right for every citizen. At the time, the Finnish government said that every citizen had the right to access a 1Mbps broadband connection, with the goal of universal access to 100 Mbps by 2015. By 2016–2017, in an updated and more modest target, the Finnish government had a stated goal to increase the minimum speed broadband connection to every business and permanent residence in the country to 10 Mbps by 2021. By the end of 2015, 76 per cent of Finns had access to a 100 Mbps mobile broadband connection and the mobile broadband network had expanded to such a degree that it was estimated that 99 per cent of the Finnish population would be covered by the end of 2017. In relation to rural and sparsely populated areas, since 2010 the Finnish state has implemented an aid project worth €130 million to improve access to high-speed broadband. Roughly 5 per cent of the Finnish population lives in a geographical area that makes up 75 per cent of the country. By the end of 2015, high-speed broadband had been made available to 70,000 residents in these areas, and the project is planned to run until the end of 2019.

Greenland

In 2014, the government of Greenland adopted its 'Digitization as a driver of growth strategy, 2014–2017', the goal of which was to increase broadband speeds in all populated areas of Greenland by 2018. According to the Arctic Council,

Greenland's strategy on telecommunications infrastructure, 'is technology neutral and sets targets for the availability of broadband services in all towns and settlements'. By 2018 all populated areas should have access to 10 Mbps and 80 per cent of the population should have access to at least 30 Mbps. At the time of the Arctic Council report, broadband services in Greenland had a household penetration of 45 per cent, but most of the connections provided were between 2 and 10 Mbps. In early 2018, Greenland realized its goal of increased access to high-speed broadband when the national provider, TELE Greenland, expanded its broadband network of 30 Mbps for downloads and 2 Mbps for uploads to 14 of Greenland's 18 towns and settlements, or 80 per cent of Greenland's population. Unlimited access to this maximum 30 Mbps speed, however, costs over US$ 150 per month. TELE Greenland aims to offer high-speed broadband to over 90 per cent of the population by 2020. According to a 2018 report the use of mobile data has exploded in recent years, and eight in ten people now have access to the Internet through their mobile devices (Tusagassiuutit, 2018).

Iceland

Iceland has a well-developed and efficient system in place for high-speed broadband access. One area where residents of more sparsely populated parts of the country are at a disadvantage is in access to high-speed fixed-line broadband, but the Icelandic government has put in place a plan to ensure that there will be (near) 100 per cent coverage of fixed-line broadband by 2020, and that almost 100 per cent of the country will have access to high-speed mobile broadband by 2022. At the moment, there is 99 per cent coverage of the country via 3G and 98 per cent coverage via 4G.

Norway

Norway's National Communications Authority (NCOM) defines broadband as a minimum service capacity greater than 4 Mbps. In 2013, NCOM set a target of 100 Mbps throughout the country. Three years later, the Norwegian government repeated that 100 Mbps broadband should be available to at least 90 per cent of households by 2020, with a long-term goal of universal 100 Mbps access. Norway increased its rural broadband coverage by 6 per cent in 2013, and by 2014 a network capable of 30 Mbps was available to just under 80 per cent of homes in the country and just over 30 per cent of rural households. In the Arctic region, broadband is available as either wireless or fibre optic networks in both Troms and Finnmark. Full 4G coverage was near completion at the end of 2017. To highlight Norway's commitment to full broadband coverage across the whole of its Arctic territory, in order to assist fisheries, the military and research, in 2018 it was announced that the government would contribute €105 million to a project run by Space Norway whereby two satellites providing high-speed broadband access will provide 24-hour coverage to latitudes above 65 degrees north. It is anticipated that the satellites will be launched in 2022 and have a 15-year lifespan.

Russia

The Ministry of Communications and Mass Media of the Russian Federation announced national broadband targets in 2012. The ministry plans to make 100 Mbps ultra-fast broadband (UFB) available to 80 per cent of Russian residents by 2018, with the ultimate goal of providing affordable and accessible broadband to 95 per cent of households by 2020. In addition, the Russian government has a goal of 100 per cent availability of mobile broadband Internet in all communities with a population above 10,000. A 2015 study by the World Bank showed that a fixed broadband service was available to 56 per cent of Russian households. Despite these claims—and as indicated by the sardonic phrase 'internyet' used by Russians living in the north to describe the quality of their access—service in the Arctic regions has been problematic. In response, in March 2018 Russian President Vladimir Putin announced a new programme to provide 'almost universal fast Internet access' in the northern region by 2024. A fibre optic communication line will be laid to all settlements with a population over 250, including in the Arctic region.

A recent news report (Kolomychenko, 2018) notes that Russia's state security agency is opposing a high-level deal for the US OneWeb satellite startup to provide Internet access to remote parts of Russia on the grounds that the project could be used to gather intelligence, thereby damaging national security. A Federal Security Service official, Vladimir Sadovnikov, stated: 'The only way to address the threats of foreign satellite networks like OneWeb, especially in the Arctic region and Far North, is to restrict their usage in Russia'. He added that Russia would prefer a partnership with non-aggressive countries such as India and China.

Sweden

Sweden has long maintained a goal of universal access to high-speed broadband. In late 2016, the Swedish government published its broadband strategy document, *A Completely Connected Sweden by 2025*. In it, the government laid out clear and precise goals for the spread of high-speed broadband access. By 2020, for example, the goal was to provide 95 per cent of all Swedish households and businesses with access to broadband at a minimum speed of 100 Mbps. By 2025, 100 per cent of Swedish households and businesses should have access to high-speed broadband. Of this 100 per cent, 98 per cent should have access to an impressive 1 Gbps, 1.9 per cent to 100 Mbps and 0.1 per cent to 30 Mbps. As with many other countries, however, there is still a rural–urban divide with regard to access to broadband. In 2016–2017, 67 per cent of households had access to broadband at a speed of 100 Mbps, while the proportion in rural and less-populated areas was 21 per cent. Interestingly, this divide is no greater in the Arctic region, where the corresponding numbers are 65 per cent and 18 per cent, respectively. Efforts have also been made to increase the spread of mobile broadband, and 77 per cent of the country now has access to mobile broadband of up to 10 Mbps, although surface coverage declines to 56 per cent when narrowed to the Swedish Arctic.

United States/Alaska

In January 2014, the US government published an *Implementation Plan for The National Strategy for the Arctic Region*, in which federal agencies were asked to assess 'the telecommunication infrastructure in the Arctic and use new technology to support improved communications in the region, including in areas of sparse population to facilitate emergency response'. In 2014, the State of Alaska set a goal of 100 Mbps connectivity for every Alaskan household by 2020. As of 2016–2017, 126 Alaskan communities had a broadband service via fibre optic cables or a fixed wireless service. In 2016, the State of Alaska Department of Commerce reported that 70 per cent of rural Alaskans and 87 per cent of non-rural Alaskans subscribed to an Internet service. Since 2011, access to a wired connection of at least 10 Mbps has increased from 74.8 per cent to 83.9 per cent of Alaskans, but large numbers remain isolated. In 2018, 161,000 people in Alaska had no access to a wired connection capable of 25 Mbps download, 237,000 had access to only one wired service provider, leaving them no options to switch, and 57,000 had no wired Internet providers available where they live.

Media, communication materialities and geopolitics

The infrastructural dimensions of communication, as exemplified by Internet access, have implications for conceptualizing geopolitics in a world where change is entangled with both ideas and materialities, including technologies and access to them. Specifically, the dialectical relationship between media and nature, which was scrutinized in early 20th-century German philosophy in respect of *the media being environments* themselves and *the environments being the media* (see Christensen and Nilsson, 2018) provides a fertile basis for a co-constructivist understanding of space and the Arctic. Taking the two simple examples of satellites and photography, including digital imagery, as technological mediums, it is possible to say that we would not have known what we know today about the Arctic (not to mention the planet as a whole) without data reception and interpretation techniques, which ultimately find their way into the everyday media as catchy illustrations of the region (Peters, 2015; Wormbs, 2013). A similar argument can be made about the availability of and access to communication networks as structuring forces. Framing does not only apply to textual meaning but to representations through imagery, and still photography and moving images of wildlife and Arctic environmental change have been defining features of journalistic coverage and media messaging, as our research results reveal. At an age where 'alternative truth-claims' that use both visual and textual representations travel far, credible journalism and the capacity of local Arctic residents to connect to communication networks that potentially speak to large audiences matter greatly.

The factors that play into journalistic and media representations of the Arctic are many. These include the political economy of institutions, journalistic norms, ideological polarization, and a market place of agendas and discourses where competing news stories are short-lived and episodic. The insights gathered from our quantitative and qualitative content analyses and interviews reveal a portrait of an Arctic that has gained global prominence and visibility due to its destabilized environment. Yet, its stereotypical representations, with their lack of adequate depth and local context, persist in international media. Another crucial dimension of media materiality is the connectivity of the Arctic region itself. Decades after digital networks became a global feature and a definer of development, digital communications are becoming a must in the Arctic primarily due to state and private interests in the region, which necessitate connectivity. This has complex reverberations for the four million residents of the Arctic in terms of becoming part of the global digital highway, including access to shaping the media discourses and the business implications of such connectivity for the big players. As Pinkerton (2013) contends, journalists are geopolitical agents who occupy a critical position between practical and geopolitical discourses, elites and the everyday, frontline events such as wars and disasters, and local and international audiences. Journalism has expanded over the past decade to include citizen accounts and digital witnessing, placing citizens as well as geopolitical agents.

An approach that highlights the complex nature of geopolitics that includes the role of the media makes visible the hierarchically ordering roles of mediation, state politics and corporate interests. On a broader level, geopolitical considerations help to pinpoint historical and contemporary geographic interplays and shifting planetary scales by way of placing centre-stage human impacts and non-human systems in the context of colonial legacies and current dynamics of subordination (Burkart and Christensen, 2013; Dalby, 2013; Elden, 2013; Moisio, 2015). What ultimately defines the geopolitics of the Arctic today is the subject of a diverse variety of publications that set out both complementary and contested visions. Classical conceptualizations of geopolitics that underscore the supremacy of hard power have sustained relevance in the face of climate change, international politics and the deeper penetration of the market economy in the region. These dynamics bring with them open future scenarios of both peace and cooperation and conflict and contestation. Nonetheless, the big state and corporate players in the Arctic have more common and pragmatic interests rather than contested interests.

Critical geopolitics epistemologically underscores political economic and geographic relations as historically and socially constructed through hard power as well as rooted social imaginaries and representations. Embedding the 'media materialities' perspective—beyond media representations—opens up ways of rethinking geopolitics in general and Arctic geopolitics in particular. Such a renewed scope accommodates both content/discourse and infrastructural dimensions through a fresh lens, and in that way complements the perspectives on dichotomous constructions of hard power versus soft power and culture versus nature. We return to these questions in Chapter 6.

Concluding remarks

Our understanding of the Arctic region is fluid, and geopolitical, economic and technological changes all have an impact on discourses. The various media play a major role in shaping these discourses and are part of making the region a new frontier for exhibitions of geopolitical power. Media content reflects the vast quantities of research and data demonstrating the impact of human activity on the global climate and the region. Other stories are neglected, however, such as those related to the rights and concerns of local and indigenous people who are disproportionately affected not only by climate change, but also by the concrete relationship between capitalism, profit, oil and climate. As the informants from two major newspapers discussed, environmental coverage writ large is interlinked with the political economy of the media, as well as global realpolitik. Moreover, while the legacy media might be doing as good a job as they can under the circumstances that define their production logic, their focus remains on news about natural disasters, tragedies and scientific breakthroughs. One way potentially to bypass the discursive power of the mainstream media is by providing access to alternative outlets and forums via local and hyper-local news outlets, social media, chat groups and video sharing. Such forums are online for the most part, but access to high-quality broadband—and thus high-speed Internet—is often sparse in the Arctic locales. The most isolated areas have been relegated to high-cost, low-quality communications. An understanding of how national governments are attempting to remedy this situation is a first step toward understanding how the future might look in the battle to tell the story of the Arctic.

References

ACIA (2005) *Arctic Climate Impact Assessment 2005*. Cambridge: Cambridge University Press.

Aldred J (2007) Q&A: Bali climate change conference. *The Guardian*, 2 January.

Alexander R and Puh L (2014) Alaska Dispatch buys Anchorage Daily News. *KTOO* 8 April. Available at: www.ktoo.org/2014/04/08/alaska-dispatch-buys-anchorage-daily-news/ (accessed 14 August 2018).

Arctic Council (2015) Iqaluit Declaration. Ninth Ministerial Meeting of the Arctic Council. 24 April 2015. Iqaluit, Yukon, Canada. Available at: https://oaarchive.arctic-council. org/handle/11374/662 (accessed 14 October 2015).

Arctic Council (2017) *Telecommunications Infrastructure in the Arctic: A Circumpolar Assessment*. Arctic Council Secretariat. Available at: https://oaarchive.arctic-council. org/handle/11374/1924 (accessed 14 November 2018).

Arctic Economic Council (2016) *Arctic broadband: Recommendations for an interconnected Arctic*. Arctic Economic Council.

Arctic Today (n.d.) About Us. Available at: www.arctictoday.com/about-us/ (accessed 14 August 2018).

Barents Observer (2018) BarentsObserver. Available at: http://wayback.archive-it. org/10184/20180313082607/http://barentsobserver.com/en (accessed 8 November 2018).

Berzina I (2015) Foreign and domestic discourse on the Russian Arctic. In: Heininen L, Exner-Pirot H and Plouffe J (eds) *Arctic Yearbook 2015*. Akureyri, Iceland: Northern Research Forum, pp. 281–295. Available at: https://arcticyearbook.com/ arctic-yearbook/2015/2015-scholarly-papers/136-foreign-and-domestic-discourse-on-the-russian-arctic (accessed 13 March 2019).

Bird M (2015) Making waves: The navy's Arctic ambition revealed. *Globe and Mail*, 4 March.

Bravo MT and Sörlin S (eds) (2002) *Narrating the Arctic: A Cultural History of Nordic Scientific Practices*. Canton, MA: Science History Publications.

Breum M (2018) How the narrative on polar bears has become a problem for Arctic environmental groups. *Arctic Today*, 21 October. Available at: www.arctictoday. com/narrative-polar-bears-become-problem-arctic-environmental-groups/ (accessed 14 November 2018).

Broadband Now (2018) Internet access in Alaska: Stats and figures. Available at: https:// broadbandnow.com/Alaska (accessed 8 November 2018).

Burkart P and Christensen M (2013) Geopolitics and the popular. *Popular Communication* 11(1): 3–6. DOI: 10.1080/15405702.2013.751853.

Bushue A (2015) *Framing of Military Activity in the Arctic on Russia Today*. Master's thesis. Uppsala University, Department of Informatics and Media. Available at: www.diva-portal.org/smash/get/diva2:824449/FULLTEXT01.pdf (accessed 13 December 2018).

Buurman T and Christensen M (2017) Governance and the changing Arctic: News framings in US newspapers from 2007 to 2015. In: *Conference on Communication and Environment (COCE)*, Leicester, 28 June–3 July 2017.

Canyon Meyer M (2010) Alaska Dispatch enterprise reporting from the last frontier. *Columbia Journalism Review*, 29 December.

Chater A and Landriault M (2016) Understanding media perceptions of the Arctic Council. In: Heininen L, Exner-Pirot H and Plouffe J (eds) *Arctic Yearbook 2016*. Akureyri, Iceland: Northern Research Forum, pp. 61–74. Available at: https:// arcticyearbook.com/arctic-yearbook/2016/2016-scholarly-papers/168-understanding-media-perceptions-of-the-arctic-council (accessed 13 March 2019).

Christensen M (2013) Arctic climate change and the media: The news story that was. In: Christensen M, Nilsson AE, and Wormbs N (eds) *Media and the Politics of Arctic Climate Change: When the Ice Breaks*. New York: Palgrave Macmillan, pp. 26–51.

Christensen M and Nilsson AE (2017) Communicating climate change and Arctic sea ice loss. *Popular Communication* 15(4): 249–268.

Christensen M and Nilsson Annika E (2018) Media, communication, and the environment in precarious times. *Journal of Communication* 68(2): 267–277. DOI: 10.1093/ joc/jqy004.

Clark P (2018) Interview by Miyase Christensen, London, September 2018.

Dalby S (2013) Rethinking geopolitics: Climate security in the Anthropocene. *Global Policy* 5(1): 1–9.

Davies W, Wright S and Van Alstine J (2017) Framing a 'Climate Change Frontier': International news media coverage surrounding natural resource development in Greenland. *Environmental Values* 26(4): 481–502. DOI: 10.3197/096327117X14976 900137368.

DeGeorges K (2017) Interview by Annika E Nilsson in connection with the Fairbanks Ministerial meeting May 2015.

DeGeorges K (2018) Interview by Miyase Christensen at the Arctic Circle Assembly 2018.

Devonne J (2014) Former ADN editor speaks about sale. In: *The Mudflats: All Things Interesting from The Upper Left Corner*. Available at: www.themudflats.net/ archives/43067 (accessed 14 August 2018).

Devyatkin DA, Suvorov RE and Sochenkov IV (2017) An information retrieval system for decision support: An Arctic-related mass media case study. *Scientific and Technical Information Processing* 44(5): 329–337. DOI: 10.3103/S0147688217050033.

Elden S (2013) Secure the volume: Vertical geopolitics and the depth of power. *Political Geography* 34: 35–51.

Feschenko V (2011) Sporov po Arktike net [No Arctic Disputes]. *Rossiyskaya Gazeta*, 21 January.

Globe and Mail (2007) Canada 2099: How global warming will change the country we live in. 27 January.

Government Offices of Sweden (2017) A completely connected Sweden by 2025: A broadband strategy. Stockholm: Government Offices of Sweden. Available at: www.government.se/information-material/2017/03/a-completely-connected-sweden-by-2025--a-broadband-strategy/ (accessed 8 November 2018).

Gritsenko D (2016) Vodka on ice? Unveiling Russian media perceptions of the Arctic. *Energy Research & Social Science* 16: 8–12. DOI: 10.1016/j.erss.2016.03.012.

Harvey F (2018) Interview by Miyase Christensen, London, 17 September 2018.

Hiebert BC (2014) *'Heroes for the Helpless': How National Print Media Reinforce Settler Dominance through their Portrayal of Food Insecurity in the Canadian Arctic*. Thesis. Available at: https://qspace.library.queensu.ca/handle/1974/8639 (accessed 26 April 2018).

IPCC (1995) *Climate Change 1995: The Science of Climate Change. Summary for Policymakers*. IPCC/UNEP/WMO.

Klimenko E, Nilsson AE and Christensen M (2019) *Narratives of conflict and cooperation in the Arctic and Russian media*. SIPRI Insights on Peace and Security. Stockholm: SIPRI.

Kolomychenko M (2018) Citing security concerns, Russia opposes US start-up promising satellite Internet to remote areas. Reuters, 25 October.

Krivoshapko Y (2016) Zimu brosilo v zhar [Winter is heating up]. *Rossiyskaya Gazeta*, 1 February.

Leonard SP (2011) Greenland's race for minerals threatens culture on the edge of existence. *The Guardian*, 6 February.

McCarthy S (2011) US panel warns on Arctic drilling. *Globe and Mail*, 11 January. Available at: www.theglobeandmail.com/report-on-business/industry-news/energy-and-resources/us-panel-warns-on-arctic-drilling/article561902/ (accessed 7 November 2018).

Medred C (2014) Alaska Dispatch and adn.com combining in expanded site. *Alaska Dispatch News*, 6 July. Available at: www.adn.com/alaska-news/article/note-readers/2014/07/07/1.

Moisio S (2015) Geopolitics/critical geopolitics. In: Agnew J et al. (eds) *The Wiley Blackwell Companion to Political Geography*. Chichester: Wiley, pp. 220–234.

Nansen F (1897) *Fram over Polarhavet. Den Norske Polarfærd 1893–1896*. Aschehoug.

Nefidova N (2014) *Environmental Public Debate: In the Context of the Arctic in Russian and Norwegian Media*. Master's thesis. University of Oslo, Center for Development and the Environment. Available at: www.duo.uio.no/handle/10852/41797 (accessed 26 April 2018).

Nicol H (2013) Natural news, state discourses & the Canadian Arctic. In: Heininen L, Exner-Pirot H and Plouffe J (eds) *Arctic Yearbook 2013*. Northern Research Forum, pp. 211–236.

Nilsen T (2015) Interview by Annika E Nilsson, Kirkenes 23 October 2015.

Nilsen T (2018a) Comment at the session 'Arctic journalism at the crossroad of technology, economy, and politics', 20 October 2018 at the 2018 Arctic Circle Assembly.

Nilsen T (2018b) Interview by Miyase Christensen, September 2018.

Nilsen T (2018c) All Russian Arctic settlements to get fast internet. *Barents Observer*, 1 March. Available at: https://thebarentsobserver.com/en/arctic/2018/03/all-small-arctic-russian-settlements-get-fast-internet (accessed 8 November 2018).

Nilsson AE and Döscher R (2013) Signals from a noisy region. In: Christensen M, Nilsson AE and Wormbs N (eds) *Media and the Politics of Arctic Climate Change: When the Ice Breaks*. New York: Palgrave Macmillan, pp. 93–113.

Nisbet MC and Scheufele DA (2009) What's next for science communication? Promising directions and lingering distractions. *American Journal of Botany* 96(10): 1767–1778. DOI: 10.3732/ajb.0900041.

Norsk Redaktørforening (n.d.) Association of Norwegian Editors. Available at: www.nored.no/Association-of-Norwegian-Editors (accessed 13 August 2018).

Northwestel (2018) High-speed Internet comes to Nunavut: Northwestel officially launches new broadband satellite network. News release. Available at: http://globenewswire.com/news-release/2018/09/17/1571913/0/en/High-speed-Internet-comes-to-Nunavut-Northwestel-officially-launches-new-broadband-satellite-network.html (accessed 8 November 2018).

Peters JD (2015) *The Marvelous Clouds: Toward a Philosophy of Elemental Media*. Chicago: University of Chicago Press.

Pincus R and Ali SH (2016) Have you been to 'The Arctic'? Frame theory and the role of media coverage in shaping Arctic discourse. *Polar Geography* 39(2): 83–97. DOI: 10.1080/1088937X.2016.1184722.

Pinkerton A (2013) Journalists. In: Dodds K, Kuus M and Sharp JP (eds) *The Ashgate Research Companion to Critical Geopolitics*. Farnham, UK; Burlington, VT: Ashgate, pp. 439–460.

Quinn E (2018) Comment at the session 'Arctic journalism at the crossroad of technology, economy, and politics', 20 October 2018 at the 2018 Arctic Circle Assembly.

Reistad HH (2016) *Norway's Arctic Conundrum: Sustainable Development in the Norwegian Media Discourse*. Master's thesis in Sustainable Development 320. Uppsala University. Dep of Earth Sciences, Uppsala, Sweden. Available at: https://uu.diva-portal.org/smash/get/diva2:1039287/FULLTEXT01.pdf (accessed 7 March 2019).

Rogoff A (2018) Interview by Miyase Christensen at the Arctic Circle Assembly 2018.

Rothrock DA, Zhang J and Yu Y (2003) The Arctic ice thickness anomaly of the 1990s: A consistent view from observations and models. *Journal of Geophysical Research: Oceans*, 108(C3).

Sergunin AA and Konyshev VN (2016) *Russia in the Arctic: Hard or Soft Power?* Stuttgart: ibidem Press.

Smolyakova T (2011) Arktika menyayet klimat [the Arctic is changing the climate]. *Rossiyskaya Gazeta*, 28 October.

Sorokina N (2007) Skhvatka za Arktiku [Fight for the Arctic]. *Rossiyskaya Gazeta*, 3 August.

Staalesen A (2018) Two new satellites to boost Norway's Arctic internet. *Barents Observer*, 27 March. Available at: https://thebarentsobserver.com/en/arctic/2018/03/two-new-satellites-boost-norways-arctic-internet (accessed 8 November 2018).

Steinberg PE, Bruun JM and Medby IA (2014) Covering Kiruna: A natural experiment in Arctic awareness. *Polar Geography* 37(4): 273–297. DOI: 10.1080/1088937X.2014.978409.

Steinberg PE, Tasch J and Gerhardt H (2015) *Contesting the Arctic: Rethinking Politics in the Circumpolar North*. London: I B Tauris & Co Ltd.

Stoddart MCJ and Smith J (2016) The endangered Arctic, the Arctic as resource frontier: Canadian news media narratives of climate change and the North. *Canadian Review of Sociology/Revue canadienne de sociologie* 53(3): 316–336. DOI: 10.1111/cars.12111.

Strand T (2015) Kilde til NRK: – Russisk etterretning ba Norge om å bringe BarentsObserver til taushet. *NRK* 3 October. Available at: www.nrk.no/norge/kilde-til-nrk_-_-russisk-etterretning-ba-norge-om-a-bringe-barentsobserver-til-taushet-1.12583998 (accessed 13 August 2018).

TeleGeography (2017) TELE Greenland expands VDSL services to 80% of population. Available at: www.telegeography.com/products/commsupdate/articles/2017/12/13/tele-greenland-expands-vdsl-services-to-80-of-population/index.html (accessed 8 November 2018).

Tjernshaugen A and Bang G (2005) *ACIA og IPCC en sammenligning av mottakelsen i amerikansk offentlighet.* No. 2005:4. Oslo: Cicero, Center for International Climate and Environmental Research. Available at: https://brage.bibsys.no/xmlui/handle/11250/191998 (accessed 7 March 2019).

Tusagassiuutit (2018) *Tusagassiuutit: en kortlægning af de grønlandske medier.* University of Greenland, Afdeling for Journalistik og Ilisimatusarfik. Available at: https://uni.gl/media/4352533/dktusagassiuutit2018rapport.pdf.

Walsh D (2011) Climate change takes a toll on cultures. *New York Times*, 27 September.

Webb T (2010) Cairn Energy fails to find enough oil off the coast of Greenland. *The Guardian*, 26 October.

Wilson Rowe E (2013) A dangerous space? Unpacking state and media discourses on the Arctic. *Polar Geography* 36(3): 232–244. DOI: 10.1080/1088937X.2012.724461.

Wilson Rowe E and Blakkisrud H (2012) *Great Power, Arctic Power: Russia's Engagement in the Circumpolar North.* NIPU Policy Brief 2: 2012. Oslo, Norway: Norwegian Institute of International Affairs. Available at: www.nupi.no/en/Publications/CRIStin-Pub/Great-Power-Arctic-Power-Russia-s-engagement-in-the-High-North (accessed 7 March 2019).

Wormbs N (2013) Eyes on the ice: Satellite remote sensing and the narrative of visualized data. In: Christensen M, Nilsson AE and Wormbs N (eds) *Media and the Politics of Arctic Climate Change: When the Ice Breaks.* New York: Palgrave Macmillan, pp. 52–69.

Wormbs N, Döscher R, Nilsson AE, et al. (2017) Bellwether, exceptionalism, and other tropes: Political co-production of Arctic climate modelling. In: Heymann M, Gramelsberger G and Mahony M (eds) *Cultures of Prediction in Atmospheric and Climate Science.* London and New York: Routledge, pp. 159–177.

Yurieva D (2007) Beskhrebetnye [Spineless]. *Rossiyskaya Gazeta*, 4 August.

Zak A (2017) Alaska Dispatch News files for bankruptcy: New publishers emerge. *Alaska Dispatch News*, 27 August.

3   A circumpolar narrative takes shape

The surge in the past decade in images and stories about the Arctic in the mainstream media may be unprecedented but narratives about the north have a much longer history. They were often developed as part of colonial ventures in the form of scientific exploration, missionary work and resource exploitation (e.g. Bravo and Sörlin, 2002; Steinberg et al., 2015). The tools for their creation were a mixture of exploration heroism, scientific classifications and maps, presence in the field and business proposals (as amply exemplified in Sörlin, 2016). Some of these narratives have been pervasive, even if the details and their prominence change over time. Many still play a role in shaping the public image of the region, and their expression in museums and exhibitions serves as a rich source of empirical material for studies of critical geopolitics. Often they provide conflicting messages that influence discussions about the goals of Arctic governance (Young and Einarsson, 2004; Cornell et al., 2016).

One of the most pervasive images of the Arctic is as a space for scientific discovery. This was initially supported by the diaries of, and news stories on, the early explorers (e.g. Nansen, 1897). Today it is further fuelled by universities and scientific organizations distributing press releases and media-savvy images to highlight scientific fieldwork and new scientific insights, often geared to a web-based media landscape. Historically, such stories often served as a source of inspiration for building a national identity. Today they still play a role when countries assert an Arctic identity. The mediation of the voyages of the research icebreaker *Oden* by the Swedish Polar Research Secretariat is just one example. Others include the voyages of the Chinese icebreaker *Xue Long* (*Snow Dragon*) and the research stations run by countries from across the world at 'the geopolitical node' of Ny Ålesund on Svalbard (Paglia, 2016). What could be perceived as a counter-narrative within the same theme of claiming a role in the region's future is the emphasis of indigenous peoples on Arctic spaces as homelands, which provides a case for special rights, not just a simple stake in a claim to be heard.

Intersecting with the scientific sovereignty narratives is a focus on the Arctic as a storehouse of resources. The resources in focus have shifted over time from fur and whale oil in the 18th and 19th centuries, to minerals and hydrocarbons in the 20th century. Less commercially tangible resources include iconic landscapes and species to be protected from people in nature reserves, climate cooling capacity and a space for tourists to experience something unique. Regardless of the type of resources, however, their values have almost always been defined from outside the region and not seldom accompanied by claims to a right to be part of decision making on the region's future. The resource narratives of today also create what might be called an Arctic paradox. One the one hand, the impacts of a warming climate are an existential threat to the region as they will change its key characteristics—how its landscapes and living conditions are shaped by ice and snow. On the other hand, the key Arctic resource most in focus in recent years has been hydrocarbon reserves, which, if used, will inevitably contribute to further emissions of greenhouse gases and thus to further climate change.

An understanding of the Arctic geopolitics of today and how these affect regional international governance requires an analysis of how this paradox has emerged and if indeed it is a paradox at all. This chapter therefore revisits two speeches made in the 1980s that in different ways have shaped today's Arctic politics, and discusses the context in which they emerged: the famous speech by the last General Secretary of the Communist Party of the Soviet Union, Mikhail Gorbachev, made in Murmansk in 1987, in which he proclaimed the Arctic a 'zone of peace'; and a speech in 1986 in which Mary Simon from the Inuit Circumpolar Conference and Robert F Keith of the Canadian Arctic Resources Committee set out an indigenous view of sustainable development. The 1980s is a conjuncture of political and industrial developments that, in addition to being the beginning of the end of the Cold War, involved high hopes related to new offshore resources, growing environmental awareness and an emerging circumpolar indigenous polity. Building on the theoretical foundation presented by Avango, Nilsson and Roberts (2013), the analysis focuses on how some actors were successful in convincing others about their particular narrative to the extent that their perspective and accompanying interests became enshrined in international norms, and specifically in shaping the current Arctic governance regime based on circumpolar political cooperation.

The Murmansk speech and the Arctic as a zone of peace

From a foreign policy perspective, the Arctic has been narrated both as a theatre of war or military conflict and as a zone of peace, according to the time and the perspective of the author. Much attention in recent years has been devoted to the remilitarization of the Arctic, including monitoring of the military capabilities of different countries (e.g. Wezeman, 2016). However, while military expansion and power capture the attention and imaginations of media reporting and scholarly commentary, the soft power of political speeches should not be overlooked. As Nye (1990) notes, soft power plays a central role in being 'able to control the political environment' and getting other countries to follow a certain agenda (p. 55). A prominent example of how a new narrative and its mediation served to shift the political environment in the circumpolar Arctic is the so-called Murmansk speech of 1987, in which Mikhail Gorbachev proclaimed the Arctic a 'zone of peace'. It was a speech that set in motion a process of the de-securitization of the far north (Åtland, 2008).

Although the speech was part of a ceremony in which Gorbachev presented 'The order of Lenin and the Gold Star' to the City of Murmansk, it was not mainly aimed at a local audience. It was live broadcast on Soviet national television and in other Eastern Bloc countries, and the content focused on international issues, such as distinct military security proposals, along with proposals related to energy, shipping, scientific cooperation, cooperation among indigenous peoples and environmental cooperation (Åtland, 2008). Its messages were immediately picked up by news media outside the Soviet Union. The *New York Times* reported on Gorbachev's suggestion that Warsaw Pact and NATO countries consider restricting military activities in the Baltic, North, Norwegian and Greenland Seas (Taubman, 1987). Its framing was congruent with contemporary US public debates on the Arctic, which had a heavy emphasis on military capacities and security interests (e.g. Young, 1985). By contrast, the *Washington Post* and the UPI press agency located the speech as part of the détente process that was ongoing at the time and also highlighted domestic themes such as Soviet price reform and putting into practice the ideas of *perestroika* (Bohlen, 1987; Mitchell, 1987).

The focus of the coverage in the Scandinavian press was different. For example, the Norwegian media included comments from the director of the Norwegian Polar Institute about a breakthrough for Arctic research, including access to Soviet knowledge about the Arctic Ocean and new business opportunities for Norwegian companies in the exploitation of offshore oil resources and onshore mining opportunities. Further media comments in the months that followed highlighted the remaining unresolved border issues between Norway and the Soviet Union, the need for a better picture of the real intentions behind the rhetoric in the speech and the need for Norway to step up its engagement on Arctic issues. The Swedish press focused on the scope for environmental cooperation and the opportunities that peaceful cooperation would open up for developing the potential of Arctic resources. There was also a note of caution that it would be crucial to maintain freedom of the seas, an area where a Swedish research expedition had run into trouble a few years before (Edmar, 2013).

The fact that Gorbachev's Murmansk speech suggested initiatives on a breadth of topics could explain part of its appeal as it allowed different countries to identify their specific interests. Furthermore, Gorbachev used the Arctic region, which at the time was not viewed as a *region* per se in a political sense, as a means for pulling this topical multiplicity together into a new narrative about cooperation. With his strategically chosen rhetorical formulation, 'speaking in Murmansk, capital of the Soviet Polar Region', he extended a Soviet narrative about the development of the Arctic, with its roots dating back to Stalin's ambitions in the 1930s (Emmerson, 2010: Ch. 2), to the entire circumpolar north. To achieve this goal of making the Soviet north a motor for economic development, security matters had to be addressed head on:

> I have had the opportunity to speak about 'our common European home' on more than one occasion. The potential of contemporary civilization could permit us to make the Arctic habitable for the benefit of the national economies

and other human interests of the near-Arctic states, for Europe and the entire international community. To achieve this, security problems that have accumulated in the area should be resolved above all.

(Gorbachev, 1987)

The rhetorical power of the speech can be contrasted with Arctic policy development in the United States in the 1980s. The Arctic Research and Policy Act of 1984 mentions many of the same issues, such as energy resources, military security, environmental concerns and the need for Arctic cooperation on research (United States, 1984), but does so without creating a new narrative in the way that Gorbachev managed to do in Murmansk by reframing the Arctic as a zone of peace. At the time, such a narrative ran counter to how the Arctic was perceived in the United States, where the focus was on strategic military issues and conflict (Nilsson, 2018). It is therefore not surprising that US media reporting placed the speech in a global frame related to the duopoly of the great powers rather than a regional Arctic context of economic development opportunities.

This global framing can also be contrasted with the more regional focus of reactions in the Scandinavian countries, where the speech was soon followed up by discussions in parliaments and in diplomatic circles. A contemporary Canadian commentary suggests that Gorbachev's initiative initially caught the foreign ministry of Canada by surprise (Purver, 1988), but there is also an account of a meeting of the foreign ministers of Canada and Norway in Tromsø on how both countries were willing to stretch out the hand of cooperation once Gorbachev's ideas became more concrete. The Norwegian foreign minister commented on the need for stable and rational development in a situation of major challenges related to security, the environment and science (Fyhn, 1987). In the Canadian and Scandinavian context, the geopolitics of the global great powers thus played less of a role.

While much of the speech is devoted to the need to reduce military tensions in the Arctic, this aim must be placed in the context of what specifically was at stake. For the Soviet Union the context was directly related to a narrative that put natural resources in focus:

Thirdly, the Soviet Union attaches much importance to peaceful cooperation in developing the resources of the North, the Arctic. Here an exchange of experience and knowledge is extremely important. Through joint efforts it could be possible to work out an overall concept of rational development of northern areas. We propose, for instance, reaching agreement on drafting an integral energy programme for the north of Europe. According to existing data, the reserves there of such energy sources as oil and gas are truly boundless. But their extraction entails immense difficulties and the need to create unique technical installations capable of withstanding the Polar elements. It would be more reasonable to pool efforts in this endeavour, which would cut both material and other outlays.

(Gorbachev, 1987)

While the policy processes that the Gorbachev speech set in motion focused on environmental and scientific cooperation, Swedish and Norwegian media commentaries were quick to pick up on the resources theme. This keen interest is not surprising, given the development of offshore hydrocarbon resource exploration in Norway at the time, including the discovery of gas in Snøhvit (Snow White) field in 1984. In Norway, the 'High North' was not yet discussed as a top priority foreign policy objective, but its importance is well illustrated by the key role played by Norway in negotiating the Barents regional cooperation (Young, 1998). Perhaps slightly more surprisingly, the resources narrative was also a good fit with the interests of Swedish political and industrial actors who saw the potential for Swedish technology exports as well as a need to assert their interests in offshore resources more generally, given the recent negotiation of the UN Convention on the Law of the Sea (UNCLOS). Offshore technology and deep-sea drilling were seen as areas with potential for Swedish industry, and political interests had already been established in the Ministry for Foreign Affairs in the early 1980s (Edmar, 2013). Sweden had acceded to the Antarctic Treaty in 1984 to assert its polar interests, and several of the articles in the Swedish press around that time appeared in the financial newspaper *Dagens Industri*. One of the headlines is telling: 'Sweden has to be in place when Antarctica's riches are eventually exploited' (*Dagens Industri*, 1983).

In terms of practical diplomacy, Gorbachev's proposal on environmental cooperation gained the most immediate attention, and eventually developed into political negotiations in the so-called Rovaniemi process, which resulted in the Arctic Environmental Protection Strategy in 1991 (Young, 1998; Tennberg, 1998). This focus built on an environmental narrative that was nascent at the time but grew into an international political discourse (Haq and Paul, 2012; for some Arctic-specific examples, see Petrov et al., 2017). It foreshadowed today's notion of the Arctic as a global environmental linchpin. The environmental focus of the Murmansk speech encompassed two quite different themes: concerns about transboundary pollution and climate change. An examination of the nascence of each of these discourses provides insights into how an environmental framing was able to gain strength from Gorbachev's initiative.

The climate discourse had its origins in the scientific community and the growing attention paid in the 1980s to global processes and viewing the Earth as a system (Miller and Edwards, 2001). Both the science and the international scientific communities had matured to a point where research into global change was being institutionalized at the international level, involving links to international political cooperation. Among the examples were international cooperation on meteorological research (Bolin, 2007) and developments in polar science following on from the scientific cooperation under the Antarctic Treaty, especially in the Scientific Committee on Antarctic Research (SCAR). At the time of the Murmansk speech, there was increasing interest in gaining access to the northern polar region, as well as tentative discussions about a forum for Arctic research cooperation. In Gorbachev's speech, these ideas materialized as concrete suggestions for an international scientific conference in Murmansk and a joint Arctic

Research Council. Through a series of negotiations, this shared interest resulted in the creation of the International Arctic Science Committee (IASC) in 1990 (Rogne et al., 2015). Two of IASC's early projects focused on the impacts of climate change in the north, with specific attention to the Barents Sea region and the Bering Sea region, both of which bordered what was then Soviet territory (Nilsson, 2007: 98–101, and references therein). The Gorbachev speech even alluded to the narrative about the Arctic as an engine for global weather systems when he called the Arctic and the North Atlantic a 'weather kitchen' of cyclones affecting all parts of the world. This attention to the role of the Arctic in the climate elsewhere was not unique to Gorbachev's speech. It was also part of Arctic policy discussions in the United States (Hickok et al., 1983) and a priority within the scientific community.

While the Murmansk speech does not dwell on pollution issues, it does call for joint international efforts on environmental protection. Pollution was a growing concern in several Arctic countries. Neighbouring Finland and Norway were worried about the transboundary impacts of emissions from the smelters on the Kola Peninsula on areas that had once been considered clear of industrial pollution (Tikkanen and Nienmelä, 1995). The risk of radioactive contamination from Soviet military activities was another major concern among its neighbours. In Sweden the impacts of organic pollutants such as PCBs and DDT in the Baltic Sea, and in Canada high levels of PCBs and DDT in the breast milk of indigenous women living far away from any sources of industrial pollution, were already on the political agenda (Downie and Fenge, 2003; Stone, 2015). Both Sweden and Canada had a strong interest in getting to grips with the situation. As Young (1998) noted, both environmental and scientific cooperation were considered areas of low politics and therefore suitable for international efforts despite the continuing military security tensions. The Murmansk speech created an opportunity to act and eventually institutionalize the Arctic environmental narrative in the Arctic Environmental Protection Strategy in relation to pollution and in IASC in relation to climate change research.

The indigenous narrative is not prominent in the Murmansk speech but the fact that cooperation among Arctic indigenous peoples is mentioned at all is significant as it shows links not just to military security issues and the interests of neighbouring countries, but also to an emerging international indigenous movement, including common links to the ever-present resource narrative.

Indigenous rights, sustainable development and the Arctic as home

In the environmental political discourse, the term sustainable development started to gain traction in the mid-1980s with the publication in 1987 of the report *Our Common Future* (World Commission on Environment and Development, 1987). This was conceived as a call for international political action and an attempt to bridge the gap between growing concern about the environment and powerful calls for economic development from the Global South. The document does not mention the Arctic but contains an extensive discussion about Antarctica. The global debate leading up to the launch of the concept of sustainable development began with the formulation of a World Conservation Strategy (IUCN, 1980), which included a section on the Arctic Ocean and a call for Arctic nations to 'map critical ecological areas (terrestrial as well as marine), draw up guidelines for their long-term management, and establish a network of protected areas to safeguard representative, unique, and critical ecosystems'.

In the late spring of 1986, a conference on the implementation of this strategy was held in Ottawa, Canada. Among the speakers were Robert F. Keith, from the Canadian Arctic Resources Committee, and Mary Simon, representing the Inuit Circumpolar Conference (ICC). This speech is noteworthy not because it caught the attention of the media—it did not—but because it articulated an Inuit notion of sustainable development that called attention to the issue of indigenous rights. It did so within a polemic directed at national governments seeking to use the north for large-scale industrial development without any attention to the need for local economic development. It was also a polemic directed at the conservation movement wanting to make decisions about Inuit land and traditional harvesting practices. Predating Gorbachev's Murmansk speech by a year and a half, the speech by Keith and Simon discussed how the new cross-cultural alliances in the process of forming could 'give new meaning to basic principles of international cooperation' and highlighted how the World Conservation Strategy could link to ongoing work within the United Nations on the right to peace and the right to development and 'might eventually integrate principles of peace with those of conservation and development' and also apply those at the regional level (Keith and Simon, 1987: 223).

The Inuit had started to organize politically across national borders in the 1970s in response to the increasing pressure to develop oil and gas resources in the Arctic (Shadian, 2014). The 1986 conference provided a platform to make their perspectives heard internationally. Some of the ideas reflected in the speech were also shared in other exchanges, such as dialogues with diplomats from the Arctic countries. One example is from June 1985, when representatives of the ICC met a delegation from the Swedish Ministry for Foreign Affairs during a visit to Greenland to discuss Inuit international collaboration and an 'Arctic Environmental Strategy' (Johnson Theutenberg, 1985). The Swedish diplomat Bo Johnson Theutenberg noted in his diary that the ICC wanted to 'give a kick' to the governments of the Nordic countries, the USA and Canada, and that 'The atmosphere is favourable for this right now. Soviet suspicion is calming down. Increased regional cooperation would be confidence-building in the Arctic region, to speak in ESK terms' (translation from Swedish original).[1] The paper was given to the Danish hosts as 'food for thought'.

The ideas on circumpolar international cooperation gained momentum after the Murmansk speech as the Rovaniemi process unfolded. However, when Leif Halonen was invited to outline the perspective of the Saami Council on early indigenous involvement in the negotiations on the Arctic Environmental Protection Strategy at the 2016 Arctic Frontiers meeting, he told how indigenous peoples were initially shut out of the negotiations. They had heard about the Rovaniemi meeting from a Danish friend of the Saami Council during an International Labour Organization conference in 1989 but no indigenous organization had been invited. The next consultative meeting was held in Yellowknife in April 1990, but indigenous organizations were not invited to that meeting either. According to Halonen (2016), 'Mary Simon, the President of the Inuit Circumpolar Conference (ICC) tried to take part at the meeting, but was thrown out'. A third consultative meeting was held in Kiruna, Sweden, in January 1991. By then the Saami Council had travelled to Alaska to coordinate with the ICC and link up with the newly formed Association of the Peoples of the North of the Soviet Union (later renamed the Russian Association of Indigenous Peoples of the North, RAIPON). Furthermore, the Swedish head of delegation and chair of the Kiruna meeting, Désirée Edmar, who had been part of the 1985 meeting with the ICC in Greenland, had become personally engaged with the role of indigenous peoples in the north (Edmar, 2013). A few days before the Kiruna meeting, the three organizations made a common statement asking Edmar for the right to participate in the formal consultative meeting, where governmental delegates had been invited to 'a reception with wine and dried reindeer meat, to create an informal arena for small talk'. As Halonen reported, 'Désirée Edmar listened to the indigenous voices . . . and even if her delegate advisor wanted [her] to turn us down, she convinced the other delegates that inviting us to the table was the right thing to do' (Halonen, 2016).

Unlike the Murmansk speech, which was conspicuously mediated to gain attention internationally, the indigenous voices in Arctic affairs gained momentum through the growing international networking among Arctic indigenous peoples, and with the help of international platforms where it was possible to insert new perspectives. The latter were especially relevant as they provided access to wider audiences than their own networks and the potential to influence international norms. Three forums are particularly relevant. One was the process that had begun with the World Conservation Strategy, in which the 1986 meeting where Keith and Simon spoke was part of a formal evaluation of how the strategy was being implemented with the aim of revising it. It was a context in which the very notion of sustainable development was being moulded and from which indigenous perspectives had previously been absent. This can be compared with how indigenous knowledge and perspectives are now enshrined in the Convention on Biodiversity and its Article 8(j) on Traditional Knowledge, Innovations and Practices.

A second process was the reformulation of ILO Convention 169 on Indigenous and Tribal Peoples, which was an explicit attempt at 'removing the assimilationist orientation of the earlier standards, and [r]ecognising the aspirations of these peoples to exercise control over their own institutions, ways of life and economic development and to maintain and develop their identities, languages and religions, within the framework of the States in which they live' (ILO, 1989). The third was the setting provided by the negotiations in the Rovaniemi process, including the growing interest of Arctic players, both state and indigenous peoples, in building a new circumpolar narrative. In building this narrative, indigenous voices asserted their rights as well as an internal perspective on sustainable development in the northern circumpolar world, as expressed in the presentation by Keith and Simon: 'The circumpolar world is also a homeland' (Keith and Simon, 1987: 210).

The homeland narrative gained further momentum as circumpolar cooperation developed. On the creation of the Arctic Council in 1996, the indigenous peoples' organizations gained status as permanent participants, not quite on a par with the member states but with much more formal influence than NGOs and other observers. The scalar political interplay involved in ensuring this status should not be underestimated. The global development of a sustainable development discourse and the re-evaluation of colonial perspectives in existing normative frameworks, such as those of the ILO, provided impetus and inspiration for Arctic indigenous peoples to set out their own ideas on sustainability and international cooperation as peoples with rights. From the other end of the scalar spectrum—local communities across the Arctic—came an urgent need for strategies to meet the increasingly large-scale industrialization of the north, linked to oil and gas as well as major hydroelectric dam projects such as those in Alta, Norway, and James Bay, Canada. In North America, the legal processes of settling indigenous land claims had gained momentum as the major mineral and hydrocarbon resources and access routes to them involved indigenous lands. The development of Arctic resources came to depend on negotiations and agreements between indigenous peoples and national governments.

In Greenland, the home rule referendum of 1979 set in motion the politics of independence from Denmark (for a detailed account of recent developments, see Breum, 2018). In the Fennoscandian countries, recognition of indigenous land rights occurred much later, and the Finnmark Act of 2005 is one of the few legal initiatives to create a new setting for making regional development decisions. Nonetheless, it is fair to say that the narrative of the Arctic as a homeland, which began to be articulated in the 1980s, has been powerfully projected and become part of the norms that influence all discussions on the Arctic (Bankes, 2004; Bankes and Koivurova, 2014).

How have the developments sketched out above been reflected in media reporting? A search of the LexisNexis database on the word 'Inuit' reveals that the Canadian national newspaper *Globe and Mail* featured reporting on Inuit claims for self-determination as early as the late 1970s, and several articles were published in the 1980s about conflicts between industrial development and indigenous self-determination and ways of life. It appears to have taken until the early 1990s (1991) for news about such developments to reach the international press, the first being a report in the British press (*The Times*) on the decision to create the Nunavut Territory. In 1992, the *Washington Post* ran a story about the Canadian government apologizing to Inuit for its resettlement policies of the 1950s. As far as media coverage goes, the word Inuit is clearly a Canadian discursive choice, linked mainly to internal Canadian issues. Searching on the word 'Eskimo'[2] reveals some stories in the *New York Times*, including a report from Greenland on Eskimo militancy:

After decades of meek submission to the white man, Eskimos are showing new signs of militancy. Increasingly organized to fight back, they are making new political and financial demands on the intruders from the south and exercising greater influence in the defense of the fragile Arctic environment.

(Border, 1982)

In a more neutral tone, the *Washington Post* reported in 1977 that:

Eskimos from across the polar northland began meeting today in this remote city 347 miles north of the Arctic Circle in the first international attempt at politically uniting nearly 100,000 Eskimos scattered across the world. The five-day gathering includes nearly 200 Eskimo representatives who flew in—Barrow has no access by road—from Greenland, Canada, Alaska and Finland for what is being called the first Inuit, or Eskimo People's Circumpolar Conference.

(Richards, 1977)

In general, however, US press coverage of the Arctic or Alaska appears to have been meagre in the 1970s and 1980s, apart from the attention on the trans-Alaskan pipeline which featured mainly oil interests and some voices from environmental NGOs.

In the Scandinavian context, Saami and environmental movement protests against the building of a hydroelectric dam in Alta caught the attention of the media due to the powerful conflicts between the Norwegian state apparatus and the emerging Saami movement. It included hunger strikes in Oslo and an attempt to blow up a bridge that was considered sabotage by the authorities and was described as 'one of the most dramatic political conflicts in Norway since World War II' (Andersen and Midttun, 1985). Judging from contemporary accounts, the attention of the Norwegian national media mainly focused on this conflict, and there was only limited interest in the broader issues of Saami lives and politics (Store, 1983). Nonetheless, the media archive database Mediearkivet shows a range of Saami issues in the reporting related to schooling, language, health and politics. Issues related to the Alta conflict were also reported in Sweden, mainly through the work of the TT press agency. Other issues in the Swedish media coverage of the 1980s were land-use conflicts in the reindeer herding areas with other activities such as forestry, tourism, mining and the building of a new road.

Concluding remarks

The building of an indigenous rights narrative in the late 1970s and the early 1980s played out in a range of political arenas in both national and international contexts. While the Alta case is an exception, where the strong conflicts appealed to the media, most political developments are characterized by a slow but steady expansion of and engagement with international institutions. These were open for negotiation at the time either because they were in the process of being established, as in the case of the Arctic Environmental Protection Strategy, or because they were in the process of reforming themselves, as in the case of the world conservation movement and the International Labour Organization through resolution ILO-169. It is noteworthy that many accounts of the initiation of circumpolar cooperation only highlight the seminal role of Gorbachev's Murmansk speech, while no attention is paid to other international political developments such as the growing importance of indigenous voices, which played a critical role in shaping today's circumpolar narrative of the Arctic as a zone of peace and cooperation.

In many accounts of Arctic international governance, the focus is on environmental narratives and how cooperation around shared environmental concerns shaped the region (e.g. Stone, 2015). It is a story based on the issues on which the regional regime that emerged from the thaw in the Cold War in the late 1980s has been quite successful. It is also a story that was shaped in an era when the focus of international relations theory was on institutions and regime formation (e.g. Young, 1998) more than geopolitical tensions. Nonetheless, regimes do not emerge without some shared interests. A look back at the 1980s shows that a key theme and goal of international Arctic politics—the Arctic as a zone of peace with sustainable development as a goal—took shape at a time when a major driver of political change was an increasing interest in the region's resources, especially oil and gas. This interest was linked to increasing the capacity for offshore extraction of hydrocarbons, which required greater international collaboration than developing onshore resources. Given the strong geopolitical interests at play in any discussion about access to natural resources, the resource narrative does not lend itself as easily to a storyline of peace and cooperation. Nonetheless, it could in fact provide at least as much explanatory power for the surge in interest in the region in the late 1980s as the environmental narrative.

Notes

1 European Security Conference, since renamed the Organization for Security and Co-operation in Europe. Translation by the author.
2 A term that many today consider archaic or offensive because of its colonial history.

References

Andersen SS and Midttun A (1985) Conflict and local mobilization: The Alta Hydropower Project. *Acta Sociologica* 28(4): 317–335. DOI: 10.1177/000169938502800402.
Åtland K (2008) Mikhail Gorbachev, the Murmansk Initiative, and the desecuritization of interstate relations in the Arctic. *Cooperation and Conflict* 43(3): 289–311. DOI: 10.1177/0010836708092838.
Avango D, Nilsson AE and Roberts P (2013) Arctic futures: Voices, resources and governance. *Polar Journal* 3(2): 431–446. DOI: https://doi.org/10.1080/21548 96X.2013.790197.
Bankes N (2004) Legal systems. In: *Arctic Human Development Report*. Akureyri: Stefansson Arctic Institute, pp. 101–118.
Bankes N and Koivurova T (2014) Legal systems. In: Larsen JN and Fondahl G (eds) *Arctic Human Development Report: Regional Processes and Global Challenges*. TemaNord 2014:567. Copenhagen: Nordic Council of Ministers, pp. 221–252.

Bohlen C (1987) Soviet urges talks between pacts; Gorbachev seeks to lessen military activity in northern waters. *Washington Post*, 2 October.

Bolin B (2007) *The History of the Science and Politics of Climate Change: The Role of the Intergovernmental Panel on Climate Change*. Cambridge: Cambridge University Press.

Border W (1982) Eskimos are showing new militancy. *New York Times*, 7 March.

Bravo MT and Sörlin S (eds) (2002) *Narrating the Arctic: A Cultural History of Nordic Scientific Practices*. Canton, MA: Science History Publications.

Breum M (2018) *Hvis Grønland river sig løs – en rejse i kongarigest sprekker*. Copenhagen: Gyldendal.

Cornell S, Downing A and Clark D (2016) Multiple Arctics: Resilience in a region of diversity and dynamism. In: Carson M and Peterson G (eds) *Arctic Resilience Report*. Stockholm: Stockholm Environment Institute and Stockholm Resilience Centre, pp. 27–61.

Dagens Industri (1983) Sverige blir medlem i Antarktis-klubben. *Dagens Industri*, 2 November.

Downie D and Fenge T (eds) (2003) *Northern Lights Against POPs*. Montreal & Kingston: McGill-Queen's University Press.

Edmar D (2013) Interview about Swedish Polar politics in the 1980s and early 1990s by Annika E Nilsson.

Emmerson C (2010) *The Future History of the Arctic*. London: Bodley Head.

Fyhn M (1987) Norskkanadisk utspill: Samarbeide i Arktis. *Aftenposten*, 10 December.

Gorbachev MS (1987) *The Speech in Murmansk: at the ceremonial meeting on the occasion of the presentation of the Order of Lenin and the Gold Star Medal to the city of Murmansk, October 1, 1987*. Moscow: Novosti Press Agency Pub. House. Available at: https://catalog.hathitrust.org/Record/006877290 (accessed 9 March 2018).

Halonen L (2016) The establishment and development of the Arctic Council and the Indigenous Peoples Secretariat. Presentation at Arctic Frontiers Plus, 25 January 2016. Tromsø, Norway.

Haq G and Paul A (2012) *Environmentalism since 1945: The Making of the Contemporary World*. Abingdon, Oxon; New York: Routledge.

Hickok DM, Weller G, Davis TN et al. (1983) *United States Arctic Science Policy*. Alaska Council of Science and Technology.

ILO (1989) Convention C169: Indigenous and Tribal Peoples Convention, 1989 (No. 169). ILO. Available at: www.ilo.org/dyn/normlex/en/f?p=NORMLEXPUB:12100:0::NO:: P12100_ILO_CODE:C169 (accessed 13 March 2018).

IUCN (1980) *World Conservation Strategy: Living Resource Conservation for Sustainable Development*. IUCN-UNEP-WWF.

Johnson Theutenberg B (1985) Dagbok från UD. Volume 5. Mitt decennium 1976–1987 som UD:s folkrättssakkunnige, Del III:III (1985–1987) och Del IV Polarambassadör – ÖB:s folkrättslige rådgivare 1987–1988. Kap 65 Maj-Juni 1985. Available at: www. theutenberg.se/pdf/Kap_65_MAJ-JUNI_1985.pdf (accessed 12 June 2013).

Keith RF and Simon M (1987) Sustainable development in the northern circumpolar north. In: Jacobs P and Munro DA (eds) *Conservation with Equity: Strategies for Sustainable Development*. Gland, Switzerland: IUCN, pp. 209–225. Available at: https://portals. iucn.org/library/node/5866 (accessed 7 March 2019).

Miller CA and Edwards PN (eds) (2001) *Changing the Atmosphere: Expert Knowledge and Environmental Governance*. Cambridge, Mass.: MIT Press.

Mitchell C (1987) Soviet leader Mikhail Gorbachev toured the Murmansk area for . . . Available at: www.upi.com/Archives/1987/10/02/Soviet-leader-Mikhail-Gorbachev-toured-the-Murmansk-area-for/3084560145600/ (accessed 9 March 2018).

Nansen F (1897) *Fram over Polarhavet. t: den norske polarfærd 1893–1896*. Aschehoug.

Nilsson AE (2007) *A Changing Arctic Climate: Science and Policy in the Arctic Climate Impact Assessment*. Department of Water and Environmental Studies, Linköping University.

Nilsson AE (2018) The United States and the making of an Arctic nation. *Polar Record*: 54(2): 95–107. DOI: 10.1017/S0032247418000219.

Nye JSJ (1990) Soft power. *Foreign Policy* 80: 153–171.

Paglia E (2016) The telecoupled Arctic: Ny-Ålesund, Svalbard as scientific and geopolitical node. In: *The Northward Course of the Anthropocene: Transformation, Temporality and Telecoupling in a Time of Environmental Crisis*. Stockholm: KTH Royal Institute of Technology PhD Dissertation History of science, technology and environment, pp. 17–29. Available at: http://kth.diva-portal.org/smash/get/diva2:881415/FULLTEXT 02.pdf (accessed 7 March 2019).

Petrov AN, BurnSilver S, Chapin FS et al. (2017) *Arctic Sustainability Research: Past, Present and Future*. London and New York: Routledge.

Purver R (1988) Arctic security: The Murmansk initiative and its impacts. *Current Research on Peace and Violence* 11(4): 147–158.

Richards B (1977) Near the people, Eskimos ponder political unity. *Washington Post*, 13 June.

Rogne O, Rachold V, Hacquebord L et al. (eds) (2015) *IASC 25 years*. IASC. Available at: https://view.joomag.com/iasc-25-years/0102946001421148178 (accessed 19 March 2018).

Shadian JM (2014) *The Politics of Arctic Sovereignty: Oil, Ice and Inuit Governance*. London and New York: Routledge.

Sörlin S (ed.) (2016) *Science, Geopolitics and Culture in the Polar Region: Norden Beyond Borders*. New York: Routledge.

Steinberg PE, Tasch J and Gerhardt H (2015) *Contesting the Arctic: Rethinking Politics in the Circumpolar North*. London: I B Tauris & Co Ltd.

Stone DP (2015) *The Changing Arctic Environment: The Arctic Messenger*. New York: Cambridge University Press.

Store A (1983) Er kjettinger og dynamitt bedre stoff? *Aftenposten*, 24 August.

Taubman P (1987) Soviet proposes Arctic peace zone. *New York Times*, 2 October.

Tennberg M (1998) *Arctic Environmental Cooperation: A Study in Governmentality*. Aldershot: Ashgate Publishing Company.

Tikkanen E and Nienmelä I (eds) (1995) *Kola Penisula Pollutants and Forest Ecosystems in Lapland*. Lapland Forest Damage Project.

United States (1984) Arctic Research and Policy Act of 1984 (amended 1990). Public Law 98-373-July 31, 1984; amended as Public Law 101-609-November 16, 1990. Available at: www.nsf.gov/geo/plr/arctic/iarpc/arc_res_pol_act.jsp (accessed 12 June 2017).

Wezeman S (2016) *Military Capabilities in the Arctic: A New Cold War in the High North?* SIPRI Background paper. Stockholm: SIPRI.

World Commission on Environment and Development (1987) *Our Common Future*. Oxford: Oxford University Press.

Young OR (1985) The age of the Arctic. *Foreign Policy* 61: 160–179.

Young OR (1998) *Creating Regimes: Arctic Accords and International Governance*. Ithaca, NY: Cornell University Press.

Young OR and Einarsson N (2004) Introduction. In: AHDR (ed.) *Arctic Human Development Report*. Akureyri, Iceland: Stefansson Arctic Institute, pp. 15–26.

4 Reconstruction and consolidation

In the early 2000s, the emphasis in discussions about the Arctic started to shift from a circumpolar regional focus to a global framing. Examples of the latter are books about globalization in relation to the Arctic (Heininen and Southcott, 2010), research efforts that take their starting point as 'The Global Arctic' (Heininen and Finger, 2017) and books that analyse the region from an explicitly global perspective (Keil and Knecht, 2017). This shift has been accompanied by tensions between local or regional and global scalar perspectives, in the scientific sphere as well as in relation to the politics of power over decision making, including arguments about the role of 'non-Arctic' actors in Arctic governance. Three tropes appear as recurring themes in this field of contestation, each with its own discursive implications: the Arctic as a bellwether of environmental change, rights over Arctic resources and the notion of the global Arctic. By delving beyond the surface of these three tropes and how they became powerful frames for discussing the circumpolar north, this chapter shows how today's debates about the region's future are rooted in earlier contestations over defining the Arctic.

A bellwether of environmental change

The Arctic has become a symbol of the impacts of environmental change. In today's media landscape, this symbolic role is especially prominent in relation to climate change, through images of calving glaciers, melting sea ice and polar bears. From a scientific perspective, the region's special sensitivity is linked to some of its physical features, such as the cold temperatures, the snow and ice that dominate the landscape and the fact that it is home to unique ecosystems. Over time, this environmental materiality has become interwoven with narratives and existential concerns that strike a chord with people who may never have visited the region.

Narratives about the Arctic environment have also been interwoven with global political discourses, and this has had various impacts on governance. In the early 1900s, a wish to preserve wild landscapes and iconic species laid the foundations for the creation of national parks in the Arctic and elsewhere. It also fuelled early international initiatives on wildlife conservation, in which the musk ox of Greenland and reindeer on Spitzbergen, among others, were the focus of a discussion about protecting species in places that were 'not owned by any sovereign power' (Conwentz, 1914). More modern efforts at species protection did not use such explicit colonial language but contemporary controversies surrounding the management of iconic Arctic species such as polar bears, whales and seals illustrate how narratives that do not originate in the region play an important role in Arctic environmental management and discussions about what knowledge and which worldviews should weigh most heavily in decision making. The longstanding controversy between Inuit and the European Union (EU) over the EU's ban on importing seal skin is one of the most illustrative examples (Sellheim, 2015). The polar bear narratives warrant a chapter of their own: starving, swimming in an Arctic Ocean without ice, displayed as ice sculptures in various shapes and forms at political meetings or as an attention-grabbing rhetorical tool, the polar bear has become an icon of Arctic environmental change. This icon has sometimes figured as an illustration in discussions about international environmental collaboration, such as the 1973 Polar Bear Treaty (e.g. Young and Osherenko, 1993), but more often as the very symbol of the threats posed by climate change.

An Arctic environmental narrative of more recent origin is pollution, especially pollution that reaches the pristine Arctic from industrial areas of the globe, such as persistent organic pollutants (POPs) and heavy metals. Concern about widespread pollution in the Arctic dates back to the 1980s and the discovery of high levels of PCBs and DDT in the breast milk and bloodstream of Inuit women in Canada who lived far away from industrial areas. The samples were initially taken to better understand background levels in order to be able to compare samples from industrial areas. When levels were found to be so high, this set in motion a major research programme that later played into Canada's interest in circumpolar Arctic cooperation and the creation of the Arctic Environmental Protection Strategy (AEPS) (Downie and Fenge, 2003; Stone, 2015). The narrative of the Arctic as a depository for global emissions of toxic chemicals gained further momentum when the AEPS published its first scientific assessment of pollution in the Arctic in 1997. This showed how persistent pollutants were able to travel on winds and in water to cold northern regions where they were taken up by, and efficiently stored in, the fat of marine mammals, such as seals, whales and polar bears, as well as the people who rely on these species as subsistence food (AMAP, 1997). Because the Inuit political movement had been launched in response to industrialization, as discussed in Chapter 3, scientists were not the only actors to provide input into this narrative. Inuit had an especially strong voice in Sheila Watt-Cloutier, who came to play a pivotal role in the negotiations on an international convention to limit the use and emissions of POPs, turning the question into one of human health and indigenous survival rather than simply an environmental concern (Downie and Fenge, 2003; Nilsson, 2012).

While the foundations for the attention paid to pollution in the Arctic and its implications were laid in scientific reports, scientific assessments do not necessarily lead to a new narrative or a policy shift (Mitchell et al., 2006). In the case of POPs in the Arctic, the impact was as much related to how the scientific findings were disseminated and woven into a highly personal story of toxic chemicals affecting vulnerable indigenous women and children. The narrative's influence is visible in the text of the Stockholm Convention on Persistent Organic Pollutants of 2001, which in its preamble acknowledges that 'the Arctic ecosystems and indigenous communities are particularly at risk because of the biomagnification of persistent organic pollutants and that contamination of their traditional foods is a public health issue' (Stockholm Convention, 2001).

The Arctic pollution narrative went beyond POPs. It had an existential dimension with the potential to reach beyond the people who were directly affected because it challenged the notion of the region as pristine. It also showed that the Arctic was not exempt from the negative impacts of the chemicalization of society, which the environmental movement had been discussing since the 1960s and the publication of *Silent Spring* (Carson, 1962). The Arctic was not as remote or isolated as previously imagined.

In recent years, the pollution narrative has faded from the public discourse, even if the issues continue to raise concerns. Its place has been taken by climate change and a narrative about the Arctic as bellwether of global change. This narrative has its origins in global scalar perspectives. It developed in its modern form in the 1970s and 1980s, with the increasing sophistication of climate models and the attempt to understand the role of ice and snow in the Earth's energy balance (Wormbs et al., 2017). The roots, however, go even further back and feature strong links between science and geopolitics. During the Cold War, the United States invested heavily in the geosciences, including a focus on meteorology and oceanography, as part of its competition with the Soviet Union (Doel, 2003). In a 1983 report that later led to the establishment of the US Arctic Research Commission, the Arctic was discussed as a 'sensitive indicator and regulator of worldwide climate change' (Hickok et al., 1983). In addition, as Chapter 3 discusses, Mikhail Gorbachev spoke about the Arctic as a weather engine in his 1987 Murmansk speech (Gorbachev, 1987). Climate change was also an issue in the early planning for circumpolar international scientific cooperation in what later became the International Arctic Science Committee (IASC), as a subject area in which a new committee for Arctic research 'could operate as a genuine international body', in contrast to issue areas where national interests were assumed to dominate (Roots and Rogne, 1987). Arctic research was also seen as important to worldwide science programmes such as the World Climate Research Programme (Roots et al., 1987). While IASC, once established, adopted climate change as one of its key priorities, it was not a prominent focus for the AEPS or when the Arctic Council was established in 1996 (Nilsson, 2007). This changed with the initiation of the Arctic Climate Impact Assessment (ACIA), which was a collaboration between IASC and two of the Arctic Council working groups: the Arctic Monitoring and Assessment Programme and Conservation of Arctic Flora and Fauna.

The ACIA has been called one of the most influential activities of the Arctic Council (Kankaanpää and Young, 2012) and also has a special role in supporting a narrative that places the Arctic in a global context. Its origin explains some of its impact. Planning for the ACIA caught the attention of the ICC, which became concerned that it would focus only on scientific impacts, and wanted to focus on people too, and also to ensure that there would be a specific policy outcome from the process (Nilsson, 2007: 110, 129). The ICC had learned from its earlier work on POPs that scientific assessments could serve as potent policy tools and was now building on this experience in crafting its involvement in the ACIA process (Watt-Cloutier, 2015). This goal was in alignment with the ambitions of the chair of the ACIA, Robert Corell, who wanted to push the international climate negotiations forward. Major efforts were also made to present the findings of the assessment in a way that could reach beyond the scientific community (Nilsson, 2007: 120 ff.).

Details of the ACIA process have been discussed elsewhere (Nilsson, 2007). Much of the wider attention paid to the scientific assessment at the time of its publication was linked to political controversy surrounding the production of a policy document. A study of media coverage of the ACIA in the United States showed that about half of all the news media attention focused on whether there would be a policy follow-up or the United States would try to prevent this (Tjernshaugen and Bang, 2005). There was also a distinct difference from the media reporting on the Intergovernmental Panel on Climate Change (IPCC) in the same period. This focused heavily on scientific controversy, which was barely visible in the ACIA coverage. The controversy over the policy document gathered momentum when Sheila Watt-Cloutier, representing the ICC, testified before the US Senate Committee on Commerce, Science and Transportation that the United States had tried to stop publication of the report (cited in Tjernshaugen and Bang, 2005).[1] She also repeated a narrative to which the ACIA had given momentum by combining the homeland metaphor with the Arctic as bellwether metaphor: 'My homeland—the Arctic—is the health barometer for the planet'.

On 30 October 2004, the *New York Times* published an article based on a leaked copy of the ACIA report, which was to be launched at the November ministerial meeting in Reykjavik, raising the stakes even further for the policy negotiations. According to the article: 'Several of the Europeans who provided parts of the report said they had done so because the Bush administration had delayed publication until after the presidential election, partly because of the political contentiousness of global warming'. The issue of delay referred to the ministerial meeting having been moved from September to November, allegedly to avoid interference with the US presidential election—an allegation denied by the Icelandic host of the meeting. The article's description of the report further illustrates how this scientific assessment was fashioned into an effective tool for communicating a specific message: 'The report is a profusely illustrated window on a region in remarkable flux, incorporating reams of scientific data as well as observations by elders from native communities around the Arctic Circle' (Revkin, 2004).

The text of the scientific report reveals an interesting development of the Arctic climate change narrative. Unlike earlier interest in the region from climate scientists, the report focused not only on the region as part of a global context but also on the local dynamics of climate change by bringing in indigenous knowledge as well as sciences such as ecology and limnology, which often base their analyses on observations in local case study areas. The report thus brought together local and global perspectives as well as both scientific and indigenous perspectives, which provided the media with a basis for stories that featured the human face of climate change (Nilsson, 2007).

The construction of this new narrative continued in two prominent scientific processes. The first was the Second International Planning Conference for Arctic Research in 2005. Along with the more traditional Arctic natural sciences of bio-physical processes, this featured sessions on vulnerability and resilience as well as the notion of sustainability science, which included community perspectives (Bowden, 2005; see also Petrov et al., 2017). The second was the International Polar Year, 2007–2008, which, in contrast to earlier polar years, had a distinctly human dimension and profile as was obvious from both the logotype and the research programme (Krupnik et al., 2011). The power of the ACIA was thus linked to its inclusion of new issues in the climate change discourse, which had previously mainly focused on questions relevant to scientists rather than the broader political community, as well as to the fact that the assessment was inte-grated into a process of articulating scientific priorities.

In the years immediately after publication of the ACIA, climate change, and by extension global change, was still mainly a narrative within the environmental politics discourse. However, it later gained power by virtue of the fact that climate change as an issue reached into multiple aspects of society, not least geopolitics, security and sovereignty. These issues were not part of the research agenda in the International Conference on Arctic Research Planning (ICARP) process but became prominent after the unexpected extent of the 2007 Arctic Ocean sea-ice minimum, which neither the ACIA nor the 2007 IPCC report predicted (for a review of sea ice and climate communication, see Christensen and Nilsson, 2017).

The 2007 and 2012 sea-ice minima, and the splurge in political and media atten-tion on the Arctic that followed, meant that the narrow understanding of global change started to take on new and much broader meanings. In 2015, for example, the notion of the Arctic as a bellwether for global change was used rhetorically in a prominent way by US President Barack Obama at the Global Leadership in the Arctic: Cooperation, Innovation, Engagement and Resilience (GLACIER) confer-ence in Alaska, which had been convened to push for a strong agreement at the UN Framework Convention on Climate Change Conference of the Parties in Paris in December 2015. In this gathering of the foreign ministers and other high represent-atives of 20 countries and the EU, Obama placed the issue of climate change at the centre of world politics by claiming that 'climate change is a trend that affects all trends—economic trends, security trends. Everything will be impacted' (Obama, 2015). The statement not only invoked a security narrative but also included a mul-tiplicity of topics and located Arctic change within a storyline that went far beyond the environment and far beyond the people living in the region.

The revival of the resource and sovereignty narratives

The Arctic is rich in resources only to the extent that its geological features and ecosystem processes can be harvested and turned into something useful (Avango et al., 2013). If it were not for knowledge, infrastructure and technology, combined with demand, the Arctic would not have attracted the attention of outsiders to the extent that it has ever since the commercial fur trade and large-scale whaling campaigns of the 1600–1800s. To guarantee that the geological features, ecosystem processes and spaces are perceived as valuable to someone who can invest the necessary time and energy to harvest them, however, the distribution of value must be predictable through rules of ownership and other types of governance arrangements. There must be rights to the resources. In the international system of states, this becomes a question of sovereignty. It is thus no coincidence that state control of northern regions followed in the wake of interest in the potential resources that Arctic ecosystems, geology and spaces provide.

Today's norms on sovereignty date back to when the modern state system was established in the 1600s, and the doctrine of freedom of the seas was formulated for areas beyond a narrow stretch along the coastlines of states in the mid-1700s. Both doctrines were challenged in the 20th century. The notion of total state sovereignty over national territory must now compete with new global norms such as those related to the responsibility to protect human rights as well as environmental values, regardless of sovereign rights. Pushing at the traditional regime from the other direction, the freedom of the sea doctrine has been circumscribed by the extension of national sovereignty further away from national coastlines. On the oceans, the new norms were legally codified in the Convention on the Law of the Sea (UNCLOS) in 1982, following several decades of negotiations.

The end of the UNCLOS negotiations coincided with another significant development for the Arctic: an acute need for new sources of fossil fuels for energy, with hopes pinned on Arctic offshore deposits. Onshore oil exploitation was already under way in the Soviet Union and Alaska, and to a lesser extent also in Canada. South of the Arctic, offshore technologies were being applied in the North Sea, and both Norway and the Soviet Union were actively exploring for Arctic offshore deposits. Norway drilled its first 'wild-cat' well in the Barents Sea in 1980 (Norwegian Petroleum Directorate, 2013: Chapter 5). In the Soviet Union, systematic explorations for offshore reserves in the Barents region began in the 1960s and increased in the 1970s. Exploratory drilling began in 1981 (AMAP, 2010: 2_148).

In addition to the increased interest in offshore petroleum, access to marine ecosystem resources was also an issue, especially for countries that relied heavily on income from fisheries. Iceland had recent experience of the so-called Cod Wars, when it had challenged earlier norms on territorial limits related to marine resources, while Norway and the Soviet Union had begun negotiations on access to and control of fish stocks in the Barents Sea, where there was no agreement about the marine border between the two countries. Furthermore, the ice-covered Arctic Ocean provided a place for the military to hide submarines carrying nuclear missiles, while scientists wanted access to learn more about how the Arctic interacted with global climate processes. In the Soviet Union, ambitions to industrialize the Arctic included a renewed interest in the potential for shipping via the Northern Sea Route (Papanin, 1978; Bulatov, 1989).

The conclusion of UNCLOS brought agreement on the rules of the game and a context in which disagreements could be negotiated and settled. It also brought most of the Arctic Ocean under the sovereign control of the coastal states, sometimes known as the Arctic 5: Canada, Denmark/Greenland, Norway, the Russian Federation (then the Soviet Union) and the United States. The United States has yet to sign UNCLOS but abides by its norms in practice, and it is clear from contemporary documents that maintaining the freedom of the seas was important in early discussions about the need for an Arctic policy (United States Arctic Policy Group, 1971). UNCLOS raised concerns also in Sweden, which lacks a coastline facing the Arctic Ocean and saw a risk of being denied access. UNCLOS made it important to assert a status as a polar nation, and Sweden achieved this by joining the Antarctic Treaty, including a prominent planting of a Swedish flag during a diplomatic visit to Antarctica in 1985. A speech from the time is illustrative:

> We have witnessed, and some of us even participated in one of the biggest revolutions in international law of our time, namely the Law of the Sea in a short time. Activities of states have become resource oriented in rather a dramatic way. To be able to fulfil all goals, present and future, pertaining to exploration and exploitation of resources governments must more than ever rely on science and technology. Science, foreign policy and international law are now related in a way heretofore unexperienced.
>
> (Johnson Theutenberg, 1985)

Sweden also established the Polar Research Secretariat in 1984 to support Swedish polar research and its role in politics.

UNCLOS determines the limits of a country's territorial sea and allows coastal states an exclusive economic zone that extends 200 nautical miles (or 370 kilometres) from their own coastlines. Within this range, each nation can engage in exploration activities and use available resources. Under Article 76 of the Treaty, it is also possible to apply to extend this zone up to around 650 kilometres from the coastline, provided that the given country's continental shelf extends that far, which can be established by mapping the seabed. In the Arctic, the coastal states have been engaged in such mapping and so far Norway, Russia and Denmark have submitted documentation to the Commission on the Limits of the Continental Shelf, which evaluates the scientific evidence and makes a recommendation. The Commission's recommendation for Norway was issued in 2009, while the documentation submitted by Denmark and Russia was still being evaluated as of 2018. In cases where claims overlap, it is up to the countries to negotiate. One example is the negotiation between Norway and Russia about the delineation in the Barents Sea, which was concluded in 2010 (Henriksen and Ulfstein, 2011). Chief among the overlapping claims still outstanding as of 2018 are Denmark's, Russia's and Canada's claims to part of the Lomonosov Ridge (IBRU, 2018).

Claims to the extended continental shelf are not directly affected by changes in the sea ice, even if more open water might make it more urgent for states to have their claims documented and evaluated. In its Article 234, however, UNCLOS also gives nations the right to impose environmental controls on ice-covered waters. With the ice extent shrinking, national control over waters also becomes a matter for dispute. In an attempt to keep up with changing environmental conditions, there has been a successful push from the Arctic Council to revise the Polar Code of the International Maritime Organization, and environmental provisions of the Polar Code were adopted in 2015 and entered into force in 2017. However, the fact that rules and regulations are in place and major powers agree does not preclude displays of geopolitical power. It is in this sense that high-profile Arctic voyages that showcase various countries' capacity to navigate Arctic waters should be viewed.

When UNCLOS was negotiated in the 1980s, no one seriously considered that the Arctic Ocean would ever have the potential to become an ice-free space. Attitudes changed following the record Arctic Ocean sea-ice minimum of 2007, which caught both researchers and political actors by surprise. In the summer of 2007 an expedition planted a titanium Russian flag on the seabed of the Arctic Ocean at the North Pole. This was followed by a heated exchange of words between political representatives of Canada and Russia (Steinberg et al., 2015: Chapter 2). Also in 2007, the IPCC presented its fourth assessment report, removing any doubt that climate change was real. It was starting to become clear that the Arctic Ocean was likely to transform in only a few decades into a more accessible space than the ice-laden seascape that had previously presented such major challenges for shipping and offshore industrial activities. Would the agreements made under UNCLOS hold up in the light of this change, ensuring that the rights of the Arctic 5 would remain intact? Was there a need for a new Arctic governance regime with a stronger role than the soft law high-level forum of the Arctic Council, which until now had mainly focused on scientific assessments?

These developments between 1980 and 2007 form the context for the Ilulissat Declaration, which was signed by the Arctic 5 in May 2008 and reaffirmed 10 years later at a second meeting in Ilulissat in May 2018. The 2008 meeting was convened by Denmark for the five coastal states to send 'a very clear political signal to everybody that we will manage the Arctic responsibly, that we have the international rules necessary and we will all abide by those rules' (Thomas Winkler, head of international law, Danish Foreign Ministry, as quoted in Borger, 2008). Moreover, the declaration was a reaction to the media attention on the Arctic 'race for resources' that had come in the wake of the Russian flag planting in the summer of 2007, as well as discussions among scholars, environmental NGOs and the EU about the need for a new Arctic governance mechanism. An account in the *New York Times* is illustrative. The reporter highlights how the Danish Foreign Minister, Per Stig Møller, 'alluded to that voyage and the media blitz that followed' and quotes him as saying 'we have hopefully, once and for all, killed all the myths of a "race to the North Pole"' (Revkin, 2008).

The Ilulissat Declaration did not break any new ground in relation to international law. It simply asserted that the signatories would abide by the existing norms of the Law of the Sea, and that the Arctic 5 would take responsibility for the orderly management of the Arctic Ocean (Dodds, 2013). Its significance was political, in that it quashed any illusions that all actors had equal rights in relation to the Arctic. According to Dodds (2013), this was important in paving the way for an opening up of the Arctic Council to new observers, such as China, which were forced to relate to this show of power based on the international framework for ocean governance that had been agreed in the 1980s. Moreover, it reasserted the status of state sovereignty as the foundational principle of modern international law, in response to various calls to treat the Arctic Ocean as a global commons. Indeed, the Ilulissat Declaration makes reference to this basic principle in its first sentence: 'By virtue of their sovereignty, sovereign rights and jurisdiction in large areas of the Arctic Ocean . . .' (Arctic Ocean Conference, 2008). Danish journalist Martin Breum also places the declaration in the context of the politics of Denmark–Greenland relations, where Denmark had a strong interest in asserting the Danish realm as a legitimate actor in relation to international law at a time of heightened political rhetoric about Greenland's independence from Denmark (Breum, 2018a).

While UNCLOS was as legally relevant before the 2008 Ilulissat meeting as it was after, it was not an issue that received much attention in Arctic discourses. For example, there was not a single mention of it in the science plans from the ICARP-II meeting in 2005, or in any Arctic Council ministerial declarations before 2008. The Arctic Climate Impact Assessment, completed in 2004, contains one brief mention in relation to fisheries, and the first report on the oil and gas assessment published in 2007 only discusses pathways and the environmental and health impacts of oil and gas, not rights to the resources as such. The only Arctic Council document where it is prominently highlighted is the 2009 Arctic Marine Shipping Assessment (Arctic Council, 2009a). However, since the meeting in Ilulissat, UNCLOS has become a more prominent part of the Arctic rhetoric. All the national Arctic strategies issued after the Ilulissat Declaration make reference to UNCLOS. Furthermore, whereas none of the previous Arctic Council ministerial declarations had mentioned it, the Law of the Sea appeared in the preamble of the 2009 Tromsø Arctic Council Ministerial Declaration: 'Recalling that an extensive legal framework applies to the Arctic Ocean including, notably, the law of the sea, and that this framework provides a solid foundation for responsible management of this ocean' (Arctic Council, 2009b).

Even actors that were highly critical of the Ilulissat meeting because it excluded other Arctic countries and indigenous peoples did not question the basic principles. One example is Iceland, which was upset about not being included in the meeting but pledged adherence to UNCLOS in its Arctic policy statement of 2011, in which high priority was given to redefining who should be included in the exclusive club of coastal states, as stated in its second principle paragraph: 'Securing Iceland's position as a coastal State within the Arctic region as regards influencing its development as well as international decisions on regional issues on the basis of legal, economic, ecological and geographical arguments' (Iceland Althingi, 2011). The first principle highlighted the Arctic Council as the 'most important consultative forum'.

A more explicit criticism was expressed in the Inuit Circumpolar Council's (ICC) Declaration of Inuit Sovereignty, which again adhered to the norms of sovereignty set out in the Ilulissat Declaration but redefined them by including a broader set of international norms than those expressed in UNCLOS (ICC, 2009). Highlighting that 'Inuit are a people', the declaration referenced the Charter of the United Nations, the International Covenant on Economic, Social and Cultural Rights, the International Covenant on Civil and Political Rights, the Vienna Declaration and Programme of Action, the Human Rights Council, the Arctic Council and the Organization of American States. Furthermore, it highlighted the special rights of indigenous peoples, as spelled out in the 2007 UN Declaration on the Rights of Indigenous Peoples (UNDRIP). It specifically raised the criticism that while referring to international law, the Ilulissat Declaration focused only on UNCLOS. In a scathing criticism of the narrow definition of sovereignty in the Ilulissat Declaration, the ICC declaration highlighted various agreements from recent years that recognized the right to self-determination, such as land claim settlements in Canada and self-government in Greenland, and concluded that: 'The issues of sovereignty and sovereign rights in the Arctic have become inextricably linked to issues of self-determination in the Arctic. Inuit and Arctic states must, therefore, work together closely and constructively to chart the future of the Arctic'.

The ICC declaration builds on the arguments set out in writing on sustainable development from 1985–86, as discussed in Chapter 3. The declaration was further developed in connection with the ACIA process, by explicitly connecting the discourse on the impacts of climate change with a discussion on indigenous rights, which included filing a petition to the Inter-American Commission on Human Rights to Oppose Climate Change Caused by the United States of America (Inuit Circumpolar Conference, 2005). This rights perspective is also elaborated on in Sheila Watt-Cloutier's book *The Right to Be Cold* (Watt-Cloutier, 2015). Furthermore, it is increasingly coming to the fore in discussions on 'stakeholders in the Arctic', in which representatives of indigenous peoples often reject the term stakeholders and instead emphasize the designation 'rights holders', which emphasizes a legal standing and status that is different from and stronger than that of other actors. This shifting international normative context has also been highlighted by academics (e.g. Bankes, 2004; Nicol, 2010; Bankes and Koivurova, 2014). The extent to which this status is recognized in practice has varied over time and depended on context, and the Ilulissat Declaration is a clear example that it is still often ignored when states see themselves as the only legitimate actors. Interestingly, at a follow-up meeting of the Arctic 5 in Chelsea, Canada, in 2009, US Secretary of State Hillary Clinton made critical remarks about excluding indigenous peoples (Dodds, 2013). Given that the United States a decade earlier had been at odds with Canada about including a strong voice for indigenous peoples in the Arctic Council, the statement shows how the narrative of indigenous rights had started to make inroads in national policy contexts too.

How is sovereignty discussed in the media? It is fairly prominent in the Canadian media: 187 articles in the *Globe and Mail* have contained both the words Arctic and sovereignty since the late 1970s (the earliest mention in LexisNexis is 1977), compared to only a handful of mentions in the *New York Times* and *The Guardian*. In the US quality press, Scott Borgerson from the US Coast Guard Academy discussed sovereignty issues in the *New York Times* following the Russian flag planting and the response of Canada's President Harper, with a focus on the Arctic being opened up to a race for resources and a call for the United States to sign UNCLOS (Borgerson, 2007). In 2013, the theme came up in writing about relations between Greenland and China, where the authors tried to calm the debate and clarify that Greenland was remaining firmly within the realm of Denmark and NATO (Breum and Chemnitz, 2013). Other articles gathered from a search of coverage about the Arctic in the *New York Times* focus only on relations between states, and the ICC on Inuit sovereignty is entirely absent from the coverage.

Why does the framing of sovereignty matter? Given the history of the Arctic as a resource frontier for colonial powers and the different views about desirable Arctic futures, sovereignty links directly to control over those resources and their role in the future of the region. Notions of sovereignty guide not only who has or should have a legitimate voice on different questions, but also the norms on what constitutes a valuable resource. Is it the geological features that provide the foundation for the mining and fossil fuel industries? Are ecosystem processes that serve as controls on the global climate system more valuable, or perhaps the local ecosystems that provide food and material, along with cultural identities and social networks? The norms that are negotiated within various governance systems guide the relative value attributed to these very different features of the environment, and shape the policies that build on those norms. Therefore, issues of sovereignty are ultimately about who has the moral and legal right to a voice in how the future is shaped. An emphasis on state sovereignty, as codified in UNCLOS and other legally binding international conventions, has given states a strong voice and standing relative to other actors and their interests, while agreements, declarations and governance arrangements that lack this legally binding foundation, such as the Arctic Council, have provided more room for other voices, including those of indigenous peoples. In this context, the Ilulissat Declaration and the ICC's declaration on Inuit sovereignty are about asserting or trying to change the rules of the game—the rules about how the negotiations should be carried out.

The global Arctic

Several states that do not have coastlines facing the Arctic Ocean and are not among the members of the Arctic Council have nonetheless been active in the region historically in research and the exploitation of its resources. When circumpolar cooperation was being formalized in the early 1990s, their role had to be defined. This was not without controversy, as several of the Arctic states wanted to maintain control. This was first illustrated in the discussions around IASC as a forum for scientific cooperation. Scientists in the United Kingdom, Germany,

Poland, France, the Netherlands and Japan did not want to be excluded as the new cooperation was being formalized, while some of the Arctic countries were highly sceptical about their role. However, science organizations in these countries were admitted as full members in 1991 (Hacquebord, 2015). In the Arctic Environmental Protection Strategy of 1991, the Federal Republic of Germany, Poland and the United Kingdom are mentioned as observers that assisted with its preparation, together with the ICC, the Nordic Saami Council, the Soviet Union Association of Small Peoples of the North, the United Nations Economic Commission for Europe, the United Nations Environment Programme and IASC. When the Arctic Council was created, the different standings of these organizations were codified. The indigenous peoples' organizations were made permanent participants, while other interested parties were able to apply for observer status. The United Kingdom, Germany, Poland and the Netherlands joined as observers in 1998 and France in 2000.

Following the record sea-ice minimum of 2007, which increased the prospects for further changes to the physical geography of the region, a second wave of actors wanted to participate in shaping the future of the Arctic. These included several Asian countries as well as the EU, where the European Parliament called for an Arctic Treaty based on the argument that the Arctic was a global commons (European Parliament, 2008; for a detailed account of the EU's Arctic engagement, see Raspotnik, 2018). In the time between the Arctic Council ministerial meetings in 2006 and 2009, China, Italy and the Republic of Korea, as well as the EU applied for observer status and became 'ad hoc observers' at lower level meetings, although not in the Arctic Council. The issue of admitting new observers remained highly controversial and the only decision taken was to complete a review of the role of observers that had already begun (Arctic Council, 2009b). The question remained deadlocked throughout the Danish chairmanship that followed, but by the time of the Nuuk ministerial meeting in May 2011, the Arctic countries had agreed to pursue a package solution to strengthening the Arctic Council that included defining the roles and responsibilities of observers and establishing a permanent secretariat. This work was to be finalized during the Swedish chairmanship (2011–2013). The Kiruna ministerial meeting, which concluded the Swedish chairmanship, received unprecedented media coverage compared to earlier Arctic Council meetings. In an analysis of 280 news stories from the Kiruna ministerial meeting in 2013, Steinberg et al. (2014) note

> how the media frame the Arctic as a region of increasing global significance—
> a region in which present-day participation is a strategic positioning for the
> future, and in which political presence holds symbolic significance for geo-
> political relations far beyond the region's latitudinal borders'.
>
> (p. 273)

Also significant was the fact that two competing narratives were in play, where the newspaper coverage from the Arctic countries tended to be characterized as 'northern protectionism' while journalism from other parts of the world emphasized 'global interconnectedness', even though it discussed similar issues such as natural resources, shipping routes, climate change and international political status.

The long-drawn-out discussion among the member states and permanent participants on the politics of new observer status rules ended in a requirement on observers not only to accept and support the objectives of the Arctic Council, but also to acknowledge the Arctic states' 'sovereignty, sovereign rights and jurisdictions in the Arctic' and agree that 'an extensive legal framework applies to the Arctic Ocean including, notably, the Law of the Sea, and that this framework provides a solid foundation for responsible management of this ocean' (Arctic Council Rules of Procedure, Annex 2). The Arctic Council members had thus put down on paper their rules of the game in a way that would help to ensure that they maintained control of the region.

According to Paglia (2016), this 'gatekeeper' role served as a strong motivation for potential observers to engage in Arctic science and other activities that would prove their legitimacy as Arctic actors. Moreover, it has shaped contemporary articulation of what the Arctic is, including expressions such as China's claim to be a 'near-Arctic state'. While several countries started to engage more with the Arctic, China's activities gained the most attention. These have included several Arctic voyages by China's icebreaking research vessel *Xue Long* (Snow Dragon), built in 1993 and upgraded in 2007 and 2013, establishing a Chinese research station on Svalbard in 2003 and keen interest in investing in opportunities for mineral prospecting in Greenland and in energy infrastructure in Iceland. One possible reason for the heightened attention on China's interest in the Arctic might be the growing geopolitical influence of China, where Chinese activities in the Arctic have come to be seen as a projection of a shifting global geopolitical landscape (Chaturvedi, 2012). Other writers, however, have analysed it as an expression of 'orientalism', or invoking fear of the East at a time when the Arctic as a space was opening up and less settled than previously (Dodds and Nuttall, 2016: Chapter 6). In this context, the new rules of procedure for the Arctic Council are described as an act of disciplining, whereby applicants must pledge adherence to the rules set by the Arctic Council (Dodds and Nuttall, 2016: citing Blank 2013).

Unlike many other countries, China was in no hurry to publish an official Arctic policy statement and its position was initially gleaned from various speeches, including occasions where Chinese officials talked about China as a near-Arctic state (SIPRI, 2012). In January 2018, however, China published its Arctic policy in both English and Mandarin in the Chinese newspaper *Xinhua* (*XinhuaNet*, 2018). Publication was immediately followed by blog commentaries on its messages. The Danish journalist Martin Breum (2018b) noted that: 'China will not rock the Arctic boat per se, but it will, as we have long seen in real life, vigorously pursue its interests through any and all legal channels', in a piece that also highlighted how Chinese investment in infrastructure had been welcomed across the Arctic. In another blog post, the academic and blogger, Mia Bennett, highlights the statement's language about China being 'a champion for the development of a community with a shared future for mankind' and the idea that China is a 'globally-minded country (implicitly contrasted with the selfish "America First")' and 'an active participant, builder and contributor in Arctic affairs who has spared no efforts to contribute its wisdom to the development of the Arctic region' (Bennett, 2018). Philip Wen of Reuters reported on China's ambition to develop a Polar Silk Road as part of the vast belt and road initiative from China to Europe and the Middle East (Wen, 2018).

As a piece of political rhetoric, China's official policy confirmed many earlier informal statements and was thus not itself new, but both its publication and the anticipation of its publication illustrate the increasing importance of officially articulating a claim to a voice in Arctic affairs. Before 2007, explicit Arctic policy statements aimed at an international audience were rare, and policies on the region were often a matter of internal affairs or connected with specific issues of international cooperation, with focus on common environmental concerns. The surge in Arctic policy statements after 2007 illustrates how the impacts of climate change and the declining sea ice have provided new momentum to the discursive shaping of the Arctic space. This flurry mirrors the activities of the late 1980s and early 1990s, when IASC and AEPS were negotiated, but with a wider geographic breadth of interest. A major difference is that the discussions are now part of a public and open debate about the region, because the web has made it easy to access and easy to disseminate statements and publications. It is part of a shifting media landscape.

Concluding remarks

In the early 2000s, narratives about the global Arctic started to seriously challenge the exclusively regional focus that Arctic international governance had fostered throughout the 1990s. The global framing stemmed initially from the natural sciences and insights about how the Arctic is connected to the rest of the world. These insights gained traction and attention from ongoing international environmental policy processes. Probably equally important was the fact that indigenous voices, speaking in unison with science, had been empowered by a growing indigenous movement to take an active part in the mediation of Arctic storylines. However, when the receding sea ice directed the world's attention to the Arctic, the global framing was no longer politically advantageous as it could be implicitly interpreted as the Arctic being everyone's space. Instead, Arctic states and indigenous peoples alike started to push a narrative focused on sovereignty, albeit with two different meanings. In this narrative, the global is limited to adherence to globally agreed norms and rules. In the norms summoned by indigenous peoples, the right to self-determination is emphasized, whereas the Arctic states have stressed rules that provide them with a privileged position vis-à-vis actors from outside the region. Somewhat counterintuitively, the geopolitical implication of the major shifts in Arctic climate conditions was, at least temporarily, the status quo. In an analysis of socio-ecological regime shifts at the circumpolar scale, Nilsson and Koivurova (2016) conclude:

While climate change is often portrayed as a driver of social change in the Arctic, it does not appear that the ongoing changes in the biophysical features of the Arctic region have rocked current circumpolar governance structures out of kilter. On the contrary, the ongoing climate-related changes, in particular sea ice decline, appear to have reinforced political commitment to existing legal structures.

(p. 179)

As long as there are strong interests at stake, however, the discursive struggle between global and national scalar perspectives and narratives is likely to continue.

Note

1 The link in Tjernshaugen and Bang (2005) (http://commerce.senate.gov/hearings/ testimony.cfm?id=1307&wit_id=3815) is no longer valid but the statement can be retrieved from www.ciel.org/Publications/McCainHearingSpeech15Sept04.pdf (accessed 19 March 2018). This record includes the following statement from Sheila Watt-Cloutier: 'I ask you to look seriously at the Arctic for solutions to the global debate on Climate Change. More specifically I ask you to look at the role your Department of State is playing in the Arctic Council's Arctic Climate Impact Assessment process. This assessment has been largely paid for by the United States, which has also provided an assessment secretariat based at the University of Alaska at Fairbanks. Bob Corell of Harvard University and the World Meteorological Institute has done a superb job of Chairing the exercise. The assessment is path-breaking and it is crucial that the world knows and understands what it says. Yet the Department of State is minimizing and undermining the effectiveness of this assessment process by refusing to allow policy recommendations to be published in a stand alone form just like the assessment itself. Yet, this is what ministers of foreign affairs directed when, in Barrow Alaska in October 2000, they approved the assessment'.

References

AMAP (1997) *Arctic Pollution Issues: A State of the Arctic Environment Report*. Oslo: Arctic Monitoring and Assessment Programme.

AMAP (2010) *Assessment 2007: Oil and Gas in the Arctic, Effects and Potential Effects. Vol. 1.* Oslo: Arctic Monitoring and Assessment Programme.

Arctic Council (2009a) Arctic Marine Shipping Assessment, 2009 Report. Available at: www.pame.is/index.php/projects/arctic-marine-shipping/amsa (accessed 7 March 2019).

Arctic Council (2009b) Tromsø Declaration on the Occasion of the Sixth Ministerial Meeting of the Arctic Council. 29 April 2009. Tromsø, Norway. Available at: https:// arctic-council.org.

Arctic Council Rules of Procedure. Annex 2. Available at: https://oaarchive.arctic-council. org/handle/11374/940 (accessed 16 December 2018).

Arctic Ocean Conference (2008) The Ilulissat Declaration. Arctic Ocean Conference, Ilulissat Greenland 27–28 May 2008.

Avango D, Nilsson AE and Roberts P (2013) Arctic futures: Voices, resources and governance. *Polar Journal* 3(2): 431–446. DOI: https://doi.org/10.1080/21548 96X.2013.790197.

Bankes N (2004) Legal systems. In: *Arctic Human Development Report*. Akureyri: Stefansson Arctic Institute, pp. 101–118.

Bankes N and Koivurova T (2014) Legal systems. In: Larsen JN and Fondahl G (eds) *Arctic Human Development Report: Regional Processes and Global Challenges*. TemaNord 2014:567. Copenhagen: Nordic Council of Ministers, pp. 221–252.

Bennett M (2018) It's official: China releases its first Arctic Policy. In: Cryopolitics. Available at: www.cryopolitics.com/2018/01/26/official-china-releases-first-arctic-policy/ (accessed 22 March 2018).

Blank SJ (2013) China's Arctic Strategy. *The Diplomat*, 20 June. Available at: https://thediplomat.com/2013/06/chinas-arctic-strategy/ (accessed 3 December 2018).

Borger J (2008) Arctic declaration denounced as territorial 'carve up'. *The Guardian*, 28 May. Available at: www.theguardian.com/environment/2008/may/29/fossilfuels.poles (accessed 3 December 2018).

Borgerson SG (2007) An ice cold war. *New York Times*, 8 August.

Bowden S (ed.) (2005) Second International Conference on Arctic Research Planning (ICARP II). The Arctic System in a Changing World. Conference Proceeding. *Copenhagen*, 1–2 November 2005. Available at: https://icarp.iasc.info/icarp-ii.

Breum M (2018a) *Hvis Grønland River Sig Løs – En Rejse i Kongarigest Sprekker*. Copenhagen: Gyldendal.

Breum M (2018b) The real question on China's new Arctic policy will be how the Arctic responds. *Arctic Today*, 31 January. Available at: www.arctictoday.com/real-question-chinas-new-arctic-policy-will-arctic-responds/ (accessed 22 March 2018).

Breum M and Chemnitz J (2013) No, Greenland does not belong to China. *New York Times*, 21 February.

Bulatov VN (1989) *KPSS – Organizator Osvoenyua Arktiki I Severnogo Morskogo Puti (1917–1980)* [CPSU: the Organizer of the Development of the Arctic and the Northern Sea Route (1917–1980)]. Moscow: MGU.

Carson R (1962) *Silent Spring*. Greenwich, CT: Fawcett Publications.

Chaturvedi S (2012) Geopolitical transformations: 'Rising' Asia and the future of the Arctic Council. In: *The Arctic Council: Its Place in the Future of Arctic Governance*. Toronto, Canada: Munk-Gordon Foundation and University of Lapland, pp. 225–260.

Christensen M and Nilsson AE (2017) Arctic sea ice and the communication of climate change. *Popular Communication* 15(4): 249–268. DOI: 10.1080/15405702.2017.1376064.

Conwentz H (1914) On national and international protection of nature. *Journal of Ecology* 2(2): 109–122. DOI: 10.2307/2255592.

Dodds K (2013) The Ilulissat Declaration (2008): The Arctic states, 'Law of the Sea', and Arctic Ocean. *SAIS Review of International Affairs* 33(2): 45–55. DOI: 10.1353/sais.2013.0018.

Dodds K and Nuttall M (2016) *The Scramble for the Poles: The Geopolitics of the Arctic and Antarctic*. Cambridge, UK; Malden, MA: Polity Press.

Doel RE (2003) Constituting the postwar earth sciences: The military's influence on the environmental sciences in the USA after 1945. *Social Studies of Science* 33(5): 635–666.

Downie D and Fenge T (eds) (2003) *Northern Lights Against POPs*. Montreal & Kingston: McGill-Queen's University Press.

European Parliament (2008) Arctic Governance. Texts adopted, Thursday, 9 October 2008 – P6_TA(2008)0474. European Parliament. Available at: www.europarl.europa.eu/sides/getDoc.do?type=TA&reference=P6-TA-2008-0474&language=EN (accessed 18 January 2018).

Gorbachev MS (1987) *The Speech in Murmansk: At the ceremonial meeting on the occasion of the presentation of the Order of Lenin and the Gold Star Medal to the City of Murmansk*, 1 October 1987. Moscow: Novosti Press Agency Publishing House. Available at: https://catalog.hathitrust.org/Record/006877290 (accessed 9 March 2018).

Hacquebord L (2015) How science organizations in the non-Arctic countries became members of IASC. In: Rogne O, Rachold V, Hacquebord L et al. (eds) *IASC 25 Years*. IASC, pp. 21–25. Available at: https://view.joomag.com/iasc-25-years/0102946001421148178 (accessed 19 March 2018).

Heininen L and Finger M (2017) The 'Global Arctic' as a new geopolitical context and method. *Journal of Borderlands Studies* 33(2): 199–202. DOI: 10.1080/08865655.2017.1315605.

Heininen L and Southcott C (2010) *Globalization and the Circumpolar North*. Fairbanks: University of Alaska Press.

Henriksen T and Ulfstein G (2011) Maritime delimitation in the Arctic: The Barents Sea Treaty. *Ocean Development & International Law* 42 (1–2): 1–21. DOI: 10.1080/00908320.2011.542389.

Hickok DM, Weller G, Davis TN et al. (1983) United States Arctic Science Policy. Alaska Council of Science and Technology.

IBRU Centre for Borders Research, Durham University (2018) Arctic maps. Available at: www.dur.ac.uk/ibru/resources/arctic/ (accessed 29 November 2018).

Iceland Althingi (2011) A parliamentary resolution on Iceland's Arctic policy. Available at: www.government.is/news/article/2011/05/05/A-Parliamentary-Resolution-on-Icelands-Arctic-Policy/ (accessed 7 March 2019).

Inuit Circumpolar Conference (2005) Inuit petition Inter-American Commission on Human Rights to oppose climate change caused by the United States of America. Press release, 7 December 2005.

Inuit Circumpolar Council (ICC) (2009) A circumpolar Inuit declaration on sovereignty in the Arctic. Inuit Circumpolar Council (ICC). Available at: www.inuitcircumpolar.com/wp-content/uploads/2019/01/declaration12x18vicechairssigned.pdf (accessed 7 March 2019).

Johnson Theutenberg B (1985) The Antarctic issue today: The Swedish viewpoint. Available at: www.theutenberg.se/pdf/Sweden_Antarctica.pdf (accessed 12 June 2013).

Kankaanpää P and Young OR (2012) The effectiveness of the Arctic Council. *Polar Research* 31(1): 17176. DOI: 10.3402/polar.v31i0.17176.

Keil K and Knecht S (eds) (2017) *Governing Arctic Change*. London: Palgrave Macmillan UK. DOI: 10.1057/978-1-137-50884-3.

Krupnik I, Allison I, Bell R et al. (2011) *Understanding Earth's Polar Challenges: International Polar Year 2007–2008*. Geneva: World Meteorological Organization.

Mitchell RB, Clark WC, Cash DW et al. (eds) (2006) *Global Environmental Assessments: Information and Influence*. Boston: MIT Press.

Nicol H (2010) Reframing sovereignty: Indigenous peoples and Arctic states. *Political Geography* 29: 78–80.

Nilsson AE (2007) A Changing Arctic Climate: Science and Policy in the Arctic Climate Impact Assessment. PhD Dissertation, Linköping University, Department of Water and Environmental Studies.

Nilsson AE (2012) Knowing the Arctic: The Arctic Council as a cognitive forerunner. In: Axworthy TS, Koivurova T and Hossain K (eds) *The Arctic Council: Its Place in the Future of Arctic Governance*. Toronto, Canada: Munk-Gordon Arctic Security Program, pp. 190–224.

Nilsson AE and Koivurova T (2016) Transformational change and regime shifts in the circumpolar Arctic. *Arctic Review on Law and Politics* 7(2): 179–195. DOI: 10.17585/arctic.v7.532.

Norwegian Petroleum Directorate (2013) Petroleum Resources on the Norwegian Continental Shelf. Available at: www.npd.no/en/publications/resource-reports/2013/ (accessed 7 March 2019).

Obama B (2015) Remarks by the President at the GLACIER Conference, Anchorage, AK. Available at: https://obamawhitehouse.archives.gov/the-press-office/2015/09/01/remarks-president-glacier-conference-anchorage-ak (accessed 21 February 2017).

Paglia E (2016) The telecoupled Arctic: Ny-Ålesund, Svalbard as scientific and geopolitical node. In: *The Northward Course of the Anthropocene: Transformation, Temporality and Telecoupling in a Time of Environmental Crisis.* Stockholm: KTH Royal Institute of Technology, PhD Dissertation, History of science, technology and environment, pp. 17–29. Available at: http://kth.diva-portal.org/smash/get/diva2:881415/FULLTEXT02.pdf.

Papanin I (1978) *Led i Plamen'* [Ice and fire]. Moscow: Politizdat.

Petrov AN, BurnSilver S, Chapin FS et al. (2017) *Arctic Sustainability Research: Past, Present and Future.* London and New York: Routledge.

Raspotnik A (2018) *The European Union and the Geopolitics of the Arctic.* Cheltenham, UK; Northampton, MA, USA: E E Elgar.

Revkin AC (2004) Big Arctic peril seen in warming. *New York Times*, 30 October.

Revkin AC (2008) 5 countries agree to talk, not compete, over the Arctic. *New York Times*, 29 May.

Roots EF and Rogne O (1987) Some points for consideration in discussion on the need for, feasibility, and possible role of an International Arctic Science Committee. Available at: https://iasc25.iasc.info/selected-historical-documents.

Roots EF, Rogne O and Taagholt J (1987) International communication and co-ordination in Arctic Science: A proposal for action. IASC. Available at: https://iasc25.iasc.info/selected-historical-documents.

Sellheim N (2015) The goals of the EU seal products trade regulation: From effectiveness to consequence. *Polar Record* 51(03): 274–289. DOI: 10.1017/S0032247414000023.

SIPRI (2012) China defines itself as a 'near-arctic state'. Available at: www.sipri.org/media/press-release/2012/china-defines-itself-near-arctic-state-says-sipri (accessed 18 January 2018).

Steinberg PE, Bruun JM and Medby IA (2014) Covering Kiruna: A natural experiment in Arctic awareness. *Polar Geography* 37(4): 273–297. DOI: 10.1080/1088937X.2014. 978409.

Steinberg PE, Tasch J and Gerhardt H (2015) *Contesting the Arctic: Rethinking Politics in the Circumpolar North.* London: I B Tauris & Co. Ltd.

Stockholm Convention on Persistent Organic Pollutants (2001) Stockholm Convention on Persistent Organic Pollutants. Available at: http://chm.pops.int/TheConvention/Overview/TextoftheConvention/tabid/2232/Default.aspx (accessed 15 October 2015).

Stone DP (2015) *The Changing Arctic Environment: The Arctic Messenger.* New York: Cambridge University Press.

Tjernshaugen A and Bang G (2005) *ACIA og IPCC en sammenligning av mottakelsen i amerikansk offentlighet.* No. 4. Oslo: Cicero, Center for International Climate and Environmental Research. Available at: www.cicero.uio.no.

United States Arctic Policy Group (1971) National Security Decision Memorandum 144. Available at: www.fas.org/irp/offdocs/nsdm-nixon/nsdm-144.pdf.

Watt-Cloutier S (2015) *The Right to Be Cold: One Woman's Story of Protecting Her Culture, the Arctic and the Whole Planet*. Toronto, Canada: Allen Lane.

Wen P (2018) A 'polar silk road' forms the core of China's first Arctic policy. *Arctic Today*, 28 January. Available at: www.arctictoday.com/polar-silk-road-forms-core-chinas-first-arctic-policy/ (accessed 22 March 2018).

Wormbs N, Döscher R, Nilsson AE et al. (2017) Bellwether, exceptionalism, and other tropes: Political co-production of Arctic climate modelling. In: Heymann M, Gramelsberger G and Mahony M (eds) *Cultures of Prediction in Atmospheric and Climate Science*. London and New York: Routledge, pp. 159–177.

XinhuaNet (2018) China's Arctic Policy. 26 January. Available at: www.xinhuanet.com/english/2018-01/26/c_136926498.htm (accessed 24 April 2018).

Young O and Osherenko G (1993) *Polar Politics: Creating International Environmental Regimes*. Ithaca and London: Cornell University Press.

5   A post-petroleum region?

'A good climate project'. This is how Norway's Minister of Petroleum and Energy, Terje Søviknes, described a new North Sea project that looks set to deliver oil for the next 50 years and generate NOK 900 billion in revenues but also 1 billion tonnes of carbon emissions into the atmosphere as the oil is burned. His logic was that the oil would be extracted using carbon-efficient technologies, including solar power. From his viewpoint, the responsibility for the emissions does not rest with Norway, but with the countries that burn the oil. When pressed about the possible moral considerations of developing new oil and gas fields, he responded: 'The climate challenge is global and must be solved globally' (Søviknes, 2018). Søviknes also expressed high hopes for further discoveries of oil and gas in Norway's Arctic sector.

Support for continuing oil and gas exploitation in the Arctic is not limited to state actors. At the 2018 Inuit Circumpolar General Assembly, Rex Rock, Chief Executive Officer of the Arctic Slope Regional Corporation, gave his 'full-throated endorsement of Arctic oil development' (Rosen, 2018b). 'You see, our region is dependent upon the economy that oil and gas development brings', he told *Arctic Today* in an article that also highlighted that the school he was speaking at, in Utqiagvik, Alaska, was, 'built with oil money, in a city with public services funded by oil and with homes that are heated by natural gas that is a byproduct of oil development'—and that Rock leads the world's richest indigenous organization.

These two episodes put the spotlight on a major Arctic dilemma: that the Arctic is entrenched in a fossil-fuelled global economy. As has been vividly illustrated in recent decades, the Arctic is also highly sensitive to climate change, and emissions from burning oil and other fossil fuels are the main culprits. The situation creates a double vulnerability: to climate change and to shifts in the demand for fossil fuels. To reduce the risk of the most far-reaching impacts of climate change, the emission of greenhouse gases, and thus both the consumption and the production of fossil fuels, must be cut not at the margins but dramatically. Such a cut will require finding alternative ways of fuelling the transport of people and goods, heating homes and producing energy-intensive materials and products. Slowly but steadily, a combination of new technical solutions and non-fossil forms of energy production is starting to transform transport and the production of goods.

However, the speed of this shift has been seriously insufficient to reduce the current pace of warming. Something more must happen to avoid dangerous climate change, which is the goal of the global climate agreement initially negotiated in the 1992 United Nations Framework Convention on Climate Change (UNFCCC) (United Nations, 1992) and further confirmed in the 2015 Paris Agreement (UNFCCC, 2015). This will require shifting patterns of travel and consumption, especially for the growing middle class worldwide, and among the high-emitting super rich. It will also require stronger political leadership than has been apparent so far (see e.g. IPCC, 2018).

It is arguable that societies should simply accept the inevitability of large-scale climate change and focus on building resilience and adaptive capacity in order to manage the impacts. It might also be possible to buy some time by reducing emissions of so-called short-lived climate forcers, such as soot (black carbon) and methane. These substances have a shorter atmospheric lifetime than carbon dioxide and may also be easier to target than taking on the global fossil fuel energy system as a whole. In the Arctic, soot is especially problematic because it darkens snow and ice so that they are less effective at reflecting the sun's energy back to space (Cavazos-Guerra et al., 2017). Unfortunately, these actions would only marginally address the overarching issue of slowing down the warming of the planet and reducing the risk of major shifts in climate, globally and in the Arctic (Arctic Council, 2013).

Because of the lack of progress in reducing greenhouse gas emissions, a growing chorus of voices is calling for cuts to the production of fossil fuels as a necessary complement to current policy instruments. This chorus includes voices from research on climate policy and energy economics (for a review, see Lazarus and van Asselt, 2018), from theorists highlighting how ingrained oil is in modern society (Urry, 2013) and in high-profile media stances such as that of *The Guardian*, which is running a dedicated campaign to keep fossil fuels in the ground (Rusbridger, 2015). The idea is that deliberately reducing the supply of coal, oil and gas would drive up the price of fossil fuel energy and serve as a complement to various efforts aimed at reducing demand.

This is not an easy direction of travel, given the powerful political and economic interests at stake. These include transnational energy companies, state interests related to energy security and maintaining national incomes as well as local interests related to job opportunities and tax revenues. It is therefore not surprising that addressing climate change by keeping fossil fuels in the ground has become a contentious topic in relation to Arctic hydrocarbon resources, as illustrated by the above quotes. For environmentalists, the need to reduce the supply of hydrocarbons has become an added argument against offshore drilling, in addition to the risks of pollution and oil spills in pristine environments that have been the focus for a long time. From a local perspective, however, the arguments do not align quite as nicely. It is easy to agree on the need to avoid pollution and oil spills that would be detrimental to everyone. It is a quite different issue to face a potential loss of revenues from oil and gas as well as stranded assets in a scenario of reduced production, regardless of whether the cause is plummeting global demand or political decisions explicitly aimed at cutting production.

In a scenario of reduced production, some communities in the Arctic will face the serious dilemma of rising costs linked to adaptation to climate change combined with the potential loss of revenue from oil and gas and stranded assets. The dip in oil prices following the oil market boom of 2009–2014 provided a sense of the coming challenge. In the United States, the dip had a major impact on jobs in Alaska (Baumeister and Kilian, 2016; Guettabi, 2016), which unlike many other oil-producing states has still not recovered from the recession induced by the oil price bust (Guettabi, 2018). In Norway, the number of jobs in the oil sector fell by 20 per cent between 2013 and 2016 (Pico, 2017). In Russia, the fall in the oil price put a dent in the national budget, which relies heavily on income from oil and gas (Sabitova and Shavaleyeva, 2015). It also brought the revitalization of the Russian Arctic town of Teriberka to a halt, as it was reliant on the development of offshore extraction in the Russian Arctic (Nilsson et al., 2019). In a scenario workshop in Bodø, Norway, in 2012, some participants saw greater challenges arising from a green scenario than from fossil fuel development, at least in the shorter term until the economy could adjust (van Oort et al., 2015).

The high stakes involved in the challenge of climate change, including the need to move society away from a fossil fuel-based energy system, make climate politics a good test case for the potential and limitations of international regional governance mechanisms. It raises many of the overarching issues of how a regional perspective relates to local, national and global initiatives, as well as the varied interests across spatial scales. It is also an area in which media attention and the mediation of political messages intersect with issues of governance and whose voices should weigh most heavily. To provide a foundation for a discussion of the role of regional governance in climate politics, with a focus on reducing the supply of fossil fuels, this chapter presents some brief background on the national interests at play. It also discusses how Arctic oil and gas figure in media reporting about the region and analyses the policy narratives as they appear in national Arctic policies and Arctic Council documents. Based on these and the material presented in Chapters 2 to 4 on the shifting media landscape and on the wider context of how Arctic governance has evolved over time, the chapter seeks to answer one of the core questions of this book: What role can regional international governance play when there is a strong need for political action but competing interests?

What is at stake?

Environmental organizations often argue that extracting oil in the Arctic is absurd because of the local environmental risks linked to exploration and exploitation, and the need to move away from fossil fuels in order to save the planet. These arguments came strongly to the fore in the widely mediated protests against the activities of Shell in the Chukchi Sea, especially after the Kulluk platform ran aground and had to be towed to Seattle harbour where it became a highly visible symbol of the risks of oil exploration (Funk, 2015; Johnson, 2015a). Such critical views from outsiders are not necessarily welcomed by actors in the Arctic. This was clear at an event organized by the Arctic Economic Council in connection with the Arctic Council Fairbanks ministerial meeting in May 2017, where a voice from the podium emphasized that Inuit, not outsiders, should make the decisions on their economic future. This stance is not based on a lack of concern about climate change and its impacts, where Arctic indigenous peoples have played a key role in highlighting the implications for the region (e.g. Watt-Cloutier et al., 2006), but rather a reaction against others claiming a right to impose decisions on a region that is moving toward increased political independence. The reception for such claims or impositions will always be coloured by past experiences of colonialism. Strong reactions are also fuelled by more recent experiences, such as the impacts on local northern economies in Canada and Greenland of the EU's ban on imports of seal skin, which was partly linked to a campaign in which Greenpeace played a major role (Sellheim, 2013).

In the mediation of the impacts of climate change in the Arctic, indigenous voices and voices from science and the environmental movement have amplified each other because of a shared interest in raising awareness of what is happening. Indigenous peoples are even brought to events to play exactly that role, where they are delegated to speak on behalf of the climate (Bjørst, 2012). In debates about reducing the extraction of fossil fuels in the Arctic, the intersection of different stakeholder interests is more complex, and there are conflicting lines of loyalties. To understand the challenges of moving toward a post-petroleum future for the Arctic, it is therefore necessary to understand the interests at stake for different Arctic actors.

Local situations vary across the Arctic and interests are closely linked to economic structures. Of critical importance is the extent to which local and state economies rely on income and jobs from the hydrocarbon sector, but also how other sectors, such as shipping and tourism, might be affected either by higher oil prices or by climate change. What are the available options and what are the risks involved? While they have the potential to be greatly affected, local communities rarely have much of a say in how global energy systems should develop. Such decisions are partly in the hands of energy corporations and energy-using industries, partly in the hands of national politics and partly in the hands of consumers spread across the world. Focusing on the political scene, national actors have a considerable interest in a cheap and secure supply of energy for industry and transport, among other things. In regions where the ground holds hydrocarbon reserves, states also have considerable power in determining the future production of fossil fuels, as they often control the use of land and sea, and possess the legislative muscle and economic policy tools to either support or constrain development. Four Arctic states—Russia, the United States, Norway and Canada—are significant players in the global fossil fuel energy market. In terms of future development, Greenland also figures in the discussion on Arctic hydrocarbon extraction.

Russia as the oil and gas giant

Russia possesses the lion's share of Arctic oil and gas resources and its production accounts for 95 per cent of total Arctic petroleum production. Moreover,

90 per cent of Russia's total oil production comes from the Arctic (Lindholt and Glomsrød, 2015). Most of this is from onshore fields in West Siberia that were initially developed in the 1960s and expanded in the late 1970s and 1980s (AMAP, 2010: 2_44). In the late 1970s, exploration also began offshore in the Barents and Kara seas, which led to the discovery of the huge Shtokman gas field in 1988.

Initially, hopes were high that development of the Shtokman field would bring much needed economic development to the Murmansk region, which had suffered severely in Russia's economic downturn in the 1990s, but various circumstances brought the project's development to a halt. First, there was the shale gas revolution, which undermined the market for liquefied natural gas (LNG) that would have made the project economically viable. Then, after Russia's annexation of Crimea, the import of essential technologies for offshore development was hit by sanctions by the west. As a result, the whole project was put on hold (Nilsson and Filimonova, 2013) and the high hopes for local and regional economic development were dashed (Nilsson et al., 2019). As of 2018, the only offshore development in active production is the Prirazlomnoye field. However, the development of onshore resources continues with a major focus on the Yamal Peninsula, where almost all of Russia's gas production takes place (Lindholt and Glomsrød, 2015). While future relations with European investors and markets remain uncertain, Russia has turned to the east and engaged with China, where access to energy resources plays a key role in an emerging Arctic-related relationship between the two states (Sørensen and Klimenko, 2017).

For Russia, Arctic oil and gas holds the key to both the national economy and regional development in parts of Russia's north. It is also a tool of geopolitics, in which the control of gas pipelines and prices figure into the relationships between Russian and former Soviet republics as well as between Russia and the EU (Högselius, 2013). The political stakes in Arctic oil and gas are thus very high, while at the same time the scale of production makes Russia an international oil and gas giant. Furthermore, projections of future production under different climate policy scenarios, that is with or without policies to enforce the Paris Agreement, indicate that most of the growth of oil and gas production will take place in the Russian Arctic (Lindholt and Glomsrød, 2015).

Norway's High North mission

Norway's role as a fossil fuel producer began in the 1960s with exploration of offshore resources in the North Sea and the claim to sovereignty over the Norwegian continental shelf, including agreements on delineations against the United Kingdom and Denmark. At the end of the 1970s, the area north of 62°N was also opened up for petroleum activities and exploration moved northward. In 2005, Norway launched a major new policy initiative in respect of its Arctic region. Based on a study of Norwegian media coverage, Jensen and Hønneland (2011) called this new focus 'The great narrative of the High North'. This initiative was in part based on foreign policy and Norway's geographical location as a NATO outpost facing its huge neighbour, Russia. However, the underlying current was linked to offshore petroleum resources in the Barents Sea on both the Norwegian and the Russian side. There was also a sense of urgency not to be left behind if Russia, with support from potential gas buyers in the United States, moved ahead. The hope was that Norway would become the supplier of offshore technology and know-how for the huge Shtokman field. Some of the documents of the time contained grand plans (e.g. Norwegian Ministry of Foreign Affairs, 2006). Jensen and Hønneland describe these efforts as a nation-focused parallel to the foreign policy-driven region-building efforts of the 1990s, which focused on building relations with neighbouring countries in the Barents region.

Norway's own Arctic resources were also being developed, and production from its major gas field (Snøhvid) came online in 2007. The context was one in which Norway's non-Arctic production was moving past its peak and the Arctic region was of special interest for maintaining the income that had been so essential to building Norway's modern economic welfare. The push to support the petroleum industry has had major impacts on governance. Most important from an international perspective was that in 2010 Norway and Russia finally settled their disputed maritime border in the Barents Sea, where control of resources is essential to any offshore development (Jensen, 2011). Nationally, an integrated management plan for the Barents Sea was established to provide a framework for addressing conflicting interests and resolving any potential disputes (Olsen et al., 2007). Hopes for regional economic development in the north have played a key role in the storyline of the High North. One example is a 2006 document, *Barents 2020*, in which the authors lament the fact that the goals of creating work and economic growth in northern Norway that were set in 1993 had not been met (Norwegian Ministry of Foreign Affairs, 2006).

Norway's hydrocarbon extraction is all offshore and the economic impacts, except in relation to national income, are localized to places with onshore activities. One example in the Arctic is Hammerfest, where Statoil has built its onshore terminal and LNG facility for the Barents Sea gas fields. For the local population, development has provided job opportunities and had ripple effects on the local economy, making Hammerfest an attractive place to live (Loe and Kelman, 2016). However, for Arctic Norway as a whole, the economic value added by the petroleum industry is less than 1 per cent, dwarfed by sectors such as health care & social work, as well as public administration and the defence sector. In 2012, the petroleum industry ranked only 22nd among the sectors listed by Glomsrød et al. (2015b: 59). Overall, the Norwegian petroleum industry can therefore be described as a national interest, where it contributes about half of the country's export income, and a local interest in the specific places that benefit economically from new job opportunities and investment, which creates hope in places that had previously been affected by economic hardship.

In a recent development, there has been increased attention in Norway on the sustainability of its focus on petroleum, leading to calls for a debate about what a post-petroleum era would look like (Dale and Kristoffersen, 2016). A study of media coverage of sustainability issues in the Arctic also shows a parallel upturn in attention on the issue of climate change and oil, which might signal that the question is becoming part of the public discussion in a way that it was not in previous media coverage of the Arctic (Reistad, 2016). Furthermore, local security narratives increasingly emphasize the potential risks of petroleum-related activities, as well as increasing tensions between national and local perspectives (Dale and Kristoffersen, 2018).

Alaska's oil economy

In the United States, supply from Alaska has formed an important part of the country's oil economy since the development of the Prudhoe Bay oil field in the 1970s, which came at a time when OPEC oil embargos had made the vulnerability of US energy supplies vividly apparent, through long lines at petrol stations and soaring petrol prices. The US Energy Information Administration agency ranked Alaska fifth among US states for crude oil production in April 2018 (US Energy Information Administration, 2018a). In 2017, the state produced approximately 5 per cent of US crude oil. Before fracking made new petroleum resources available further south, it was even more important, producing 17 per cent in 2005. Most of this oil comes from the fields in North Slope, Alaska (for an overview, see AMAP, 2010: Chapter 2.4). In the decades after the large-scale development of Alaska's oil began in the 1970s, the fields were of major national interest from an energy security point of view. This spurred major infrastructure investment, not least in building the trans-Alaska pipeline, and played a key role in the development of a national Arctic policy (Nilsson, 2018b). It also transformed Alaska's economy into one that is heavily dependent on income from the petroleum industry, with ups and downs that track the global market (Cole and Cravez, 2004; Guettabi, 2016, 2018).

For Alaska, revenues from oil play a central role in the state's economy, both for the public sector budget and for individuals who do not pay income tax but instead receive some of the dividends. In 2012, the oil and gas sectors accounted for 19.3 per cent of the economic value added by different sectors, surpassed only by the category 'public administration and defense' (Lindholt and Glomsrød, 2015: 38). Alaska is also a major energy user, ranked third among states in energy consumption per capita (US Energy Information Administration, 2018b). It is therefore no surprise that Alaskan politics have been pro-oil rather than focused on environmental concerns. This has included laments over former President Barack Obama's attempt to place limits on the areas in which exploration can take place and applause for President Donald J. Trump's efforts to ensure that as much of Alaska as possible is open to the oil business (Nilsson, 2018b). However, there are some signs that the hegemony of the pro-petroleum discourse in Alaska is starting to be challenged. The state is being hit hard by the impacts of climate change, and in the summer of 2018 Alaska's climate action team, appointed by the Governor, drafted a 45-page action plan that included suggestions for a system of carbon pricing, partly to offset missing income from dwindling oil revenues and partly to start to address and mitigate the effects of climate change (Rosen, 2018a). However, any such shifts are likely to meet resistance, as was quickly demonstrated in an opinion piece in *Arctic Today* just days after the action plan made the news (Cole, 2018). What happens in Alaska will also depend on federal politics, and on how energy companies assess the costs and opportunities linked to new drilling ventures, which in turn will relate to market analyses, operating costs and risks, and exploration results. How investors are affected by the mediated public image of energy companies might also play a role, which environmental NGOs will try to exploit. One example is how the press covered Obama's high-level GLACIER conference in August 2015, in preparation for the Paris climate conference later the same year. Journalists' interests were heavily focused on Shell's decision to retreat from drilling, and some in the media framed the withdrawal as an environmentalist victory (e.g. Krauss and Reed, 2015), in contrast to the victory of the oil industry only a few months earlier when Obama gave conditional approval for drilling off the Arctic coast (Davenport, 2015). The debate within Alaska about job losses was given less coverage in major international media, as were other issues that probably featured in the decision, such as environmental regulations, negative exploration results, low oil prices and high operating costs.

Canada

The Canadian Arctic contains major oil and gas reserves but these are much less developed than in Alaska (AMAP, 2010: Chapter 2.4). Oil and gas are also less important to the overall economy, and the sector accounts for less than 4 per cent of economic value added by different sectors in the region, compared to mining which is 17 per cent (Lindholt and Glomsrød, 2015: 44). The lack of development can be explained by the major challenges linked to the huge distances and lack of infrastructure, which would require major investment that is unlikely while resources that are cheaper to extract are available further south, such as the Alberta oil sands. In recent years, production in the Arctic and the value it adds to the economy have been in decline, mainly because the fields are starting to reach the end of their lifespan (Lindholt and Glomsrød, 2015: 45). Furthermore, in 2016 Prime Minister of Canada Justin Trudeau issued a statement on Arctic climate leadership together with US President Barack Obama (Prime Minister of Canada, 2016). However, like the debate in Alaska, Trudeau was unpopular in the north among those who might stand to benefit from hydrocarbon development, who pointed out that they had not been consulted about the ban beforehand (Van Dusen, 2016).

In terms of national media coverage of the Canadian Arctic in the *Globe and Mail*, the focus on oil does not dominate as much as it does in the United States and oil does not seem to have driven media attention in recent years. In the 1970s and 1980s, however, hopes were high for oil riches and these were prominent in the media, including headlines such as 'The Great Arctic Energy Rush' in *Business Weekly* on 4 January 1983. Such hopes were inspired by development in Alaska, and led to major Canadian public investment in geological surveys of the north, as documented in detail by economic historian Paul

Ward, who also highlights the high hopes connected with narratives about the destiny of Canada as a nation (Warde, 2018).

Greenland

Greenland is not an oil producer. There are considerable undiscovered offshore oil and gas resources (Bird et al., 2008) and some potential drilling sites have been located by seismic surveying, but explorative drilling has so far not resulted in any viable extraction. A number of companies have pulled out of their engagement in recent years. A scenario study assessment of future Arctic oil and gas production predicts that production is unlikely to start before 2040, after which there could be rapid growth in gas production (Lindholt and Glomsrød, 2015). Greenland's political interest in oil and gas is very much linked to hopes of improving the national economy and ultimately becoming economically independent of Denmark. However, future development will depend as much on the level of interest from foreign companies as on the government's interest in issuing licences. Environmental NGOs also claim to have had a role in stopping further prospecting (e.g. Carrell and van der Zee, 2010).

Media narratives on Arctic oil and gas

Drilling for oil in the Arctic is a risky business that requires major commitment from governments as well as from energy companies. Not only are the costs high for operating in harsh environments far from markets, but the consequences of an oil spill would create public relations risks that environmental NGOs will be eager to exploit. In such a context, narratives that create high expectations about financial or political gains play an important role in encouraging investment, as Paul Warde showed in his study of Canada in the 1970s and 1980s (Warde, 2018). Expectations raised can relate to: financial gains from a business point of view; grand narratives about nation building, as in the case of Canada but also Norway and Greenland; national energy security, a key issue for the United States in the 1970s and 1980s; and geopolitical power. Russia's use of its gas exports is a prime example of the latter. Narratives need fuel, and the media can play a pivotal role in strengthening or challenging certain ideas. It is therefore useful to take a closer look at how Arctic oil and gas resources are discussed in various media.

One of the few longitudinal studies of media coverage of the Arctic—focused on the Canadian press—shows that already in the 1970s 'it was the potential for energy resources, liquid natural gas and oil that riveted the media's attention' (Nicol, 2013: 216). Nicol notes that while the media focused on the potential of oil and gas, some alarms were already being sounded about the impacts of oil spills on fragile Arctic environments—and these made it onto the news agenda. Over the following decades, various other discourses became more apparent in the media's coverage of the Arctic, but in absolute numbers a framing centred on resources and economic development maintained a strong position in Canada's media coverage.

Our own work has focused on the coverage of Arctic oil in the *New York Times* and *The Guardian*, complemented by insights from studies of how the Arctic has been framed in Russian and Norwegian media and some examples from Sweden, which does not have any oil resources of its own.

In the *New York Times*, early attention on the Arctic was closely linked to the discovery of the Prudhoe Bay oil field in Alaska in 1968 and the subsequent building of the trans-Alaska pipeline, and the newspaper featured regular updates on political and legal processes. Looking at more recent coverage in articles from 2007–2015, a substantial part of the paper's interest in the region was still closely linked to oil. In fact, the increase in the frequency of the word 'oil' in 2009 followed by a drop in 2014 appears to mirror a similar increase and decline in the frequency of the word 'Arctic'. As is apparent from Figure 2.2, the drop also followed the fall in the oil price. The use of words such as 'climate' or 'ice' increases only later, with a peak in 2012 when the Arctic sea ice receded to an unprecedented extent for a second time in six years.

A closer look at the articles in the *New York Times* that mention the word 'oil' shows that in 2007, the emphasis was heavily on business opportunities with an eye on the soaring oil price. There is a tone of great optimism that includes outlooks on both Norway and Russia. One example is a major article about how the Snøhvid field was set to supply gas to US homes (Mouawad, 2007). In 2015, after the oil business was hit by a major drop in the global market price, the tone was much more sombre. Social issues came to the fore, as well as the curse of being dependent on oil incomes (Johnson, 2015b). The morality of extracting oil and gas in the Arctic, given the risk of pollution and the impacts of greenhouse gas emissions on the climate, also featured prominently, along with the politics of allowing exploration in the National Wildlife Reserve. One example is how the journalistic coverage highlighted the contradiction in Obama's stance, when he called the world's attention to the impacts of climate change in the Arctic while at the same time allowing offshore drilling. The story of Shell's Arctic ventures and the company's eventual withdrawal from the Arctic also featured in the *New York Times* (Krauss and Reed, 2015).

Oil is also one of the most important topics in *The Guardian*'s coverage of the Arctic in the period 2007–2015, but a quantitative snapshot of how often the word oil is used finds that attention on the Arctic does not appear to be as tightly linked to oil as it was in the *New York Times*. The most interesting feature in the patterns of shifts over time is the sharp increase in the use of both words, oil and climate, between 2014 and 2015. A closer analysis of the articles that mention the word 'oil' in *The Guardian*'s coverage of the Arctic in 2015 reveals that this attention was partly connected to a number of news events that were also covered by the *New York Times*, especially Shell's activities in the Arctic, including first the approval to drill by President Obama and later Shell's decision to abandon its activities in the Chukchi Sea and withdraw from the region. This reporting included drawing attention to campaigns by environmental NGOs as well as reporting on the political debate in the USA and the business risks involved in Arctic petroleum activities.

Specific to *The Guardian* is an explicit editorial campaign to push for fossil fuels to be kept in the ground in order to protect the climate, with several in-depth articles about the rationale and the implications for the petroleum industry. The campaign was not specifically focused on the Arctic but included enough mentions to affect the amount of coverage about oil in a sample of articles that cover the Arctic. One example is how Obama's stance on drilling and climate was used as an illustration:

> Barack Obama has the same problem. During a television interview last year, he confessed that 'We're not going to be able to burn it all'. So why, he was asked, has his government been encouraging ever more exploration and extraction of fossil fuels? His administration has opened up marine oil exploration from Florida to Delaware—in waters that were formally off-limits. It has increased the number of leases sold for drilling on federal lands and, most incongruously, rushed through the process that might, by the end of this month, enable Shell to prospect in the highly vulnerable Arctic waters of the Chukchi Sea.
>
> (Monbiot, 2015)

A closer look at the articles that include the word 'oil' reveals that in 2007, *The Guardian*'s attention on the Arctic and oil focused on geopolitical and governance issues related to rights over this 'oil-rich' region, including the infamous planting of the Russian flag on the seabed at the North Pole, within a race-for-resources framing. The Russian oil boom, including its geopolitical context, also drew journalistic attention. In contrast to the 2015 coverage, the issue of climate change was almost absent from the 2007 articles that mentioned 'oil'. Even if it appears to have been a more important topic overall, it was apparently treated as a separate issue. A steep rise in articles that included the three words 'Arctic', 'climate' and 'oil' appears to have taken place in 2015, concurrent with *The Guardian*'s explicit focus on climate change and keeping hydrocarbon resources in the ground. In coverage of the Arctic in the *New York Times*, there was also an increase in the number of mentions of oil and climate in the same article in 2015, but not to the same extent as in *The Guardian*.

Also present in *The Guardian*'s Arctic coverage but not as dominant was the longstanding discourse on drilling versus protection of the fragile Arctic environment, where the focus was more on local impacts, especially related to the risk of accidents, and the unique features of the threatened environments. It is notable that the geographic focus within the Arctic was almost exclusively on oil activities in the United States, with only a few articles about Russia and Norway.

Russian sources of information about the Arctic naturally have a different focus, but the emphasis on hydrocarbon resources would seem to be a common denominator. A large quantitative study aimed at mapping the media situation in the Russian Far North and the Arctic Zone using algorithms for automatic topic detection highlights hydrocarbon production in the Yamal-Nenets autonomous *okrug*, or district, and offshore in the Arctic Ocean as major topics. The issue of climate change was mentioned only in relation to scientific expeditions (Devyatkin et al., 2017). In a study comparing Russian and Norwegian media, Nefidova (2014) notes that the media debate was very much governed by the views of official political spokespeople, notably President Putin. Nefidova (2014: 83) found that 'the resource discourse' is one on which the Russian public keeps a 'voluntary or enforced silence'. This also held true when western environmental NGOs sought to engage with Russian oil and gas ventures, such as in the *Arctic Sunrise* 'incident' in 2013 when Greenpeace tried to climb on to the Prirazlomnaya drilling platform to halt its activities. The activists were arrested, and this was followed by a court process in which they were first accused of piracy, but this was later changed to hooliganism (Nefidova, 2014). A study of how the Arctic was covered in the official newspaper *Rossiyskaya Gazeta* between 2007 and 2016 shows how this event was initially treated as a security matter, raising risks related to threats of terrorism, but that the tone later shifted to a stronger focus on the geopolitical implications of oil and gas in the Arctic (Klimenko et al., 2019).

With regard to the overall discussion of oil and gas, the Klimenko study found that the tone was initially optimistic, with a focus on the potential to develop the resources of the Arctic shelf. Arctic offshore resources were described as a guarantee of maintaining high levels of oil production in Russia once the traditional areas of resource extraction, primarily in Western Siberia, had been depleted. The Arctic was often referred to as a 'resource base for the 21st century', quoting a famous speech by then President Medvedev (Aleksey, 2008). Between 2010 and 2013, the focus was on how to turn the Arctic, and the Arctic continental shelf specifically, into a major source of hydrocarbons and the obstacles that had to be overcome. Problems of geological prospecting, a lack of technological solutions for developing offshore deposits, and the need to stimulate national companies and attract foreign partners were the major focus.

From time to time, voices appeared to question the necessity, urgency and efficiency of Arctic shelf development. However, in contrast to the discussions in *The Guardian*, environmental concerns were not discussed as a reason to postpone tapping into the Arctic's resources. The focus was instead on questioning the competitiveness of the Arctic projects compared to other potential sources of oil extraction, such as tar sands (Dolgopolov et al., 2011). Examples of similar questions increased in connection with the closure of the Shtokman project in 2012. However, discussion of cooperation with foreign companies and the creation of favourable tax conditions for working on the Arctic shelf continued until western sanctions were imposed following Russia's annexation of Crimea, which radically changed the situation. In 2014, just months after production began at Prirazlomnaya, the impact of western sanctions became a topic of media attention related to Arctic oil and gas, resulting in a more cautious tone about its economic potential and feasibility given the absence of international partners. At the beginning of 2015, an opinion piece suggested a pause in the development of Arctic projects: 'Should we, under such conditions, force development of the shelf or the Arctic Ocean? Why, despite the great importance of this region to Russia, not take a pause in the development of the Arctic oil and gas fields?'

(Primakov, 2015). However, later in 2015–2016 the discussion returned to ways to develop the Arctic in spite of sanctions, and the absence of technology and foreign partners, rather than postponing or pausing development.

A totally different perspective is provided in the independent newspaper, *Novaya Gazeta*, where the need for Arctic resource development is repeatedly questioned. For instance, *Novaya Gazeta* called for less optimism from the government regarding the melting of the Arctic ice (Zakharov, 2009) and raised questions about the ecological risks related to the Prirazlomnaya rig in the Pechora sea (*Novaya Gazeta*, 2011), as well as the uncompetitive nature of the resource projects on the Arctic shelf (Dokuchaev, 2012). It also often criticized the approach of the Russian government to developing the Arctic, calling attention to its inefficiency and lack of regard for the environment in the region (Latynina, 2008).

In the Swedish and Norwegian press, 'oil' appeared as a subject in a handful of articles in the mid-1970s and another handful in the early to mid-1980s, connected to the potential for oil extraction on Svalbard and developments in the Russian Arctic. In the few articles that appeared before 1990, the focus ranged from industrial potential to environmental concerns and polar research. In Norway, attention on the Arctic surged in 2005, in connection with the launch of the High North Strategy which was intended to move northern Norway into a golden age based on oil and fisheries as the basis for a regional economic boom. A study of the Norwegian media discourse since 2007 (focused on the years 2012 and 2015 and the national newspaper *Aftenposten*) shows that oil was an important topic in the Arctic coverage, where articles about resource demands and extraction dominated other topics such as climate change and environmental issues (Reistad, 2016: 21). The attention on resource demand and extraction seems to have been connected to changes in the oil price, discussions about the need to extract the oil in the Arctic, the Paris Agreement and the so-called race for resources in the Arctic. Like *The Guardian*, there is an increasing focus on both oil and climate in the same article in 2015. Based on her Norwegian results, Reistad speculates that this might signal a shift that could potentially 'lead to these topics being treated together also in the political discourse' (Reistad, 2016: 41). This twist is further supported by a study of Russian and Norwegian media coverage of environmental issues and the Arctic in 2013–2014, where Nefidova observes that the Norwegian media used climate change as an argument against exploitation of the Arctic ecosystem, and describes this as a sign of a moral discourse and 'a new imaginative barrier . . . for the proponents of the resource-based economy' that is characteristic of the two countries' engagement in the Arctic (Nefidova, 2014: 125).

Given the above observations on how the media has discussed the Arctic and oil, there may have been a shift in the general discourse, whereby the economic hydrocarbon resource narrative is starting to be seriously challenged by an oil/climate/environment narrative. It is difficult to tell how enduring this shift will be, given that the empirical material only covers the period to 2015 when the Paris Agreement was moving climate change high up the international agenda at about the same time as the oil business was starting to feel the squeeze from falling oil prices. This combination may have made new extraction ventures in the Arctic less attractive from a business point of view and thus provided space for an alternative narrative to emerge. Such a window can easily close as the market shifts. The quotes at the beginning of this chapter indicate that oil optimism had returned to the political discourse by 2018, without necessarily linking plans for extraction in the Arctic to any moral obligation related to mitigating climate change.

The policy narrative

How do Arctic policy narratives compare with the media coverage? One way to get a sense of the dominant policy concerns is to examine national Arctic strategies. By 2013, all the Arctic states had published strategy documents that highlighted their political priorities, and some countries have since published updated versions or more detailed implementation plans. A closer look at the strategies launched in 2007–2013 finds both differences from and similarities to the media coverage. The most striking difference in the words used in these documents is the heavy emphasis on cooperation, which is largely absent from the media coverage, and that relatively less attention is paid to oil. The words 'climate' and 'environment' figure prominently as does 'research'. A similar emphasis on 'cooperation', and on 'climate' and 'environment', is also apparent in the biannual ministerial declaration by the Arctic Council. The attention on oil is also less prominent here, and exclusively focused on issues related to oil pollution, or in recent years on the efforts to coordinate the prevention of oil spills and pollution. However, a closer reading of the Arctic strategies, the ministerial declaration of 2013 and a vision document presented by the Arctic Council in 2013 reveals a shift in emphasis toward supporting economic development, in which cooperation becomes a tool for making the Arctic a 'safe operating space for business' (Nilsson, 2018a). While attention to climate change features prominently, there are no signs of the moral discourse in relation to the extraction of fossil fuel resources that highlights the need to keep hydrocarbons in the ground.

Looking at the activities of the Arctic Council and its working groups, a discussion of the conflicting interests at the intersection between reducing emissions of greenhouse gases and developing Arctic oil and gas resources is also absent. This is hardly surprising, given the strong national interest in oil and gas, but is, nonetheless, striking. The Arctic Council has carried out a major assessment of the impacts of oil and gas activities in the Arctic (AMAP, 2010). It was highly ambitious in mapping both the environmental and the social impacts of such activities, but explicitly excludes discussion of the connection between fossil fuels and climate change (p. 1_2). International cooperation in the Arctic Council excludes some issues that are linked to strong national interests, such as military security. It also tends to avoid or be careful about issues where strong national economic interests are at stake, such as fisheries and whaling. The Arctic Council came late to the issue of climate change but since it has been on the agenda, climate change impact assessments have played a prominent role in its work and in directing attention to the sensitivity of the region in terms of both local environments and livelihoods, and in relation to the global climate (ACIA, 2005; AMAP, 2011, 2017c). There have also been considerable efforts to map emissions of short-lived climate forcers, especially soot, and to get these under control. However, when it comes to emissions of greenhouse gases and the links to the extraction of fossil fuels in the Arctic, the Arctic Council has been silent (Nilsson and Meek, 2016).

Looking at how various media have discussed climate politics in relation to the Arctic Council, a study of how different governance bodies are covered in the media showed that black carbon emerges as a regional concern, but that the focus on the mitigation of Arctic warming mainly occurred within a global framing. In the coverage of the UNFCCC, the Arctic primarily featured as the signifier of climate change rather than in relation to specific regional-local concerns (Buurman and Christensen, 2017).

A challenge for circumpolar international governance

The impacts of global climate change in the Arctic are undeniable and will constitute a major long-term challenge to sustainable development in the region, directly through impacts on livelihoods, infrastructure and everyday life, and indirectly as the need for costly adaptation measures diverts resources from other investments. It is also beyond doubt that curbing further climate change will require dramatically reduced emissions of greenhouse gases from the use of fossil fuels. To reach the goal of the 2015 Paris Agreement—to keep warming as a global average below 1.5 or 2 °C—will require considerably more effort to reduce emissions of greenhouse gases than has been put in place so far (IPCC, 2018). Moreover, studies of the pathways to such a goal show that policies aimed at reducing the use of fossil fuels will not be sufficient, and that these will have to be complemented with cutting down on the extraction of hydrocarbons (Lazarus and van Asselt, 2018). According to one estimate, if all the oil, gas and coal from existing extraction sites is burned, the global temperature rise would exceed the 2 °C limit (Kartha et al., 2018). In the circumpolar north, this global average could lead to at least twice this rise in temperature due to Arctic amplification. If additional oil and gas fields are explored and developed, more fossil fuels will reach the market and an emerging view from research on the topic is that such additional supplies would make it impossible to reach the goals of the Paris Agreement (Lazarus and van Asselt, 2018), and thus also make it impossible to slow the pace of climate change in the Arctic.

As is repeatedly highlighted in many recent accounts of the Arctic, a large share of global unexplored hydrocarbon reserves is to be found in the region, mainly under the continental shelf (Bird et al., 2008). Based on the strong focus on oil in media reporting, Arctic oil and gas also appears to be an important driver of the surge in interest in the region in the past decade. New opportunities for mining and shipping are other factors, but one of the major attention grabbers from a media point of view has been the potential for major new oil and gas resources. This is not a surprise, given that Arctic oil and gas epitomize a range of different concerns: business interests from the fossil fuel industry, geopolitical considerations related to energy security, the wish for local economic development and jobs, and concerns about the impacts of industrial activities in fragile Arctic environments. Stories that fit neatly into recurring narratives of the north, such as those related to a race for resources or geopolitical conflicts, are especially attractive from a media point of view, as are spectacular events that symbolize the clash between incompatible concerns, such as environmentalists' protesting against the activities of oil companies. More complex stories of how oil and gas have provided financial power to indigenous peoples are more difficult to fit into mediatized events, but are starting to emerge, as are stories about the downside of being dependent on income from oil (Johnson, 2015b; MacFarquhar, 2015; see also Dale et al., 2018).

While media coverage may influence decision making, it is far from being the only factor. In regional international governance, it might not even be the major factor. In relation to Arctic oil and gas, the Arctic Council has framed the issue quite differently, as a matter of making oil and gas development compatible with good environmental management, with no attention paid to the role of fossil fuels in global climate change. It is as if the Arctic Council has been able to separate the impacts of climate change in the Arctic from a major causative factor that lies within reach for decision makers in the Arctic countries—further exploration of hydrocarbon resources with the aim of putting additional supplies of fossil fuels on the market.

At one level, this lack of attention on the negative role of Arctic fossil fuel supplies in combating climate change could appear inevitable—national economic and security interests are simply too strong in relation to the perceived risks related to climate change. It therefore appears that the potential for regional international governance to take on long-term sustainability challenges that collide with short-term national interests is very limited. The perception of the risks related to climate change might, of course, change and tip the balance, but given the extent of the impacts already observed and the vast knowledge available about the dramatic future impacts of Arctic climate change, both globally and locally, such a shift in perceptions does not seem likely any time soon. An alternative pathway—although still challenging—would be to focus on reducing the strong national and economic interests at stake on the supply side. Is there some way to make it attractive to keep fossil fuels in the ground for those who have the necessary economic and political power to make such decisions?

Based on developments in the Arctic over the past decade, such a scenario is most likely to be guided mainly by how the global market performs and whether the profitability of extraction declines to a situation in which new exploration ventures are no longer feasible. Such calculations are beyond the direct control of the Arctic countries. However, specific political decisions about what areas are to be opened up for exploration and public investment can also play a role, not least in the Arctic where both costs and risks are high. Such decisions mainly rest with national-level actors. A regional international governance body is unlikely to have any direct influence on either of these potential pathways to a post-petroleum future. This lack of formal or genuine capacity to influence raises the question of what role, if any, a regional international governance body might play in keeping fossil fuels in the ground. These questions are not relevant only to the Arctic, but analysing the specific context of the Arctic case could provide some insights into the political challenges involved in a shift to supply-side climate policies. Given that the issues are global in scope, it is also relevant to ask why the Arctic should take a lead and be at the forefront of a major transformation of the global energy system.

Kartha et al. (2018) has elaborated a framework for determining whose carbon should be kept in the ground based on equity considerations. They argue that an equitable distribution of the costs of decarbonizing the economy will be essential to halt the extraction of coal, oil and gas, and that due attention needs to be paid to creating a new foundation for replacing the livelihoods and revenues of those who are currently economically dependent on such activities. With the right mechanisms in place, they believe that the challenges are not insurmountable: 'In principle, societies could undertake a decarbonization transition in which they anticipate the transitional disruption, and cooperate and contribute fairly to minimize and alleviate it' (Kartha et al., 2018: abstract). Further elaborating and implementing such a framework would also be a way of avoiding a situation in which market forces determine whose carbon should not be extracted, in which the most expensive extraction is likely to cease first as the price of oil drops, for example, because of declining demand. In such a situation, Arctic communities could be particularly vulnerable because the economic base is often narrow and their remoteness from global markets and harsh operating conditions make Arctic oil and gas relatively more expensive compared to other sources, especially in places where the necessary infrastructure has not already been developed.

The ethical considerations in designing a supply-side policy would, according to Kartha et al., involve an 'extractor pays principle'. This is based on the insight that the total amount of hydrocarbons that can be extracted is finite if agreed climate goals are to be met. Following the principle of common but differentiated responsibilities and respective capabilities outlined in the UNFCCC, countries that have already extracted fossil fuels would have a greater obligation to contribute to a transition away from extraction. However, the agreed reference to 'capabilities' also means taking account of the respective capacities of different societies to bear any burdens involved in curbing extraction and transitioning to a low carbon world (Kartha et al., 2018), which in their view means economic, institutional and physical capacity. At the national level, it is easy to argue that the Arctic countries rank highly in a global comparison of capacity. This is also true at the international regional level, where the Arctic Council provides essential institutional capacity. Locally and at the subnational level, however, the situation varies considerably. Nonetheless, there is no reason to believe that an 'extractor pays' principle would automatically prioritize all potential extractive activities in the Arctic over some other parts of the world.

Thus far, ethical principles have played only a limited role in decisions about hydrocarbon extraction. National economic and security interests appear to rule, along with global supply and demand, which are in turn influenced by new technologies such as those used in the shale gas revolution. Ethical principles are unlikely to trump either of these drivers, even if they are enshrined in international conventions. Given the weak history of the Arctic Council in addressing the social and environmental costs of extracting fossil fuels and the strong national interests at stake, the international regional body is unlikely to take a lead on supply-side climate policy.

A politically more feasible way forward would be to focus on increasing the capacity to manage a future in which the extraction of fossil fuels in the Arctic is no longer economically viable. Such a situation could come about through declining global oil prices and an accompanying lack of interest from oil and gas companies, regardless of the political decisions made at the national level. It could also be fostered by a stronger focus in global climate politics on supply-side policies. Lazarus and van Asselt (2018) discuss the potential for various economic instruments and regulatory approaches. The latter path might create a more predictable situation for oil- and gas-dependent economies in the Arctic than the former, and may thus be easier in terms of adaptation in both local and national contexts.

A focus on increased adaptive capacity is also a context in which the Arctic Council could play an important role because of its mission to foster collaboration across national borders. It fits well with insights from recent assessments of adaptation action, which initially focused on adaptation to climate change but have also highlighted narrow economic structures as critical aspects of vulnerability (AMAP, 2017a, 2017b). The Council could also build on the knowledge base from circumpolar assessments of human development (Larsen and Fondahl, 2014), resilience (Arctic Council, 2016) and Arctic economies (Glomsrød et al., 2015a). Most importantly, building adaptive capacity does not challenge national interests in the same way as a focus on national resources. It might, therefore, not be as divisive and controversial as raising the supply of fossil fuels as an issue for international regional collaboration.

Given the current political situation both internationally and in some countries, it might even be politically counterproductive to focus on fossil fuel supply in the Arctic Council, as this could threaten international cooperation in its current form. The potential gains from a focus on adaptive capacity would include that, if it leads to improvements on the ground, it could make people in the Arctic and Arctic countries less dependent on income from oil and gas and thus less vulnerable to a shift away from fossil fuels globally, which is essential to reducing the pace and magnitude of climate change. In the longer run, it might also make those Arctic countries and communities that are dependent on fossil fuel extraction better able to see supply-side climate policies as an opportunity rather than a threat.

Concluding remarks

In international media, the Arctic is often placed in the context of global or national narratives, and the local and regional are made news pegs for bigger stories. Reporting on oil and gas is no exception. Indeed, national interests in Arctic fossil fuel resources have played a major role in the media narratives of the past decade or more. Recent reporting does highlight some of the local challenges of being vulnerable to falls in the oil price, but does not always provide much space for local agency. From a local perspective, oil can been seen both as providing economic security and welfare and as a source of insecurity, as highlighted in this chapter as well as in a new study of northern Norway, Alaska and Greenland after the 2014 drop in oil prices (Dale et al., 2018). These discussions about the future of petroleum reveal tensions along both spatial and temporal dimensions. In such tension fields, it becomes essential to control the narrative. While governance of the oil and gas sector remains in the strong grip of national interests, national narratives are challenged by the increasing pressure to take global climate change concerns into consideration as well as from local desires for self-determination in a region with a history of both colonialism and vulnerability to the economic volatility of global energy markets.

As for regional governance, strong national interests appear to have kept political discussion of limiting the production of fossil fuels off the agenda of circumpolar political cooperation. However, the Arctic Council has the potential to navigate a shift toward a post-petroleum future by supporting increased local capacity for dealing with the rapid and dramatic shifts in the global energy market that are essential if future increases in global temperature are to be kept below dangerous levels. Such support might, at the moment, provide the only feasible way forward in a difficult political situation with strong national and economic interests at stake, combined with a dire need for international cooperation and to create connections between global and local scalar perspectives. What the political geography of the Arctic region will look like in the possible future scenarios of post-petroleum societies and what strategies will be followed in rescaling governance remain to be seen.

References

ACIA (2005) *Arctic Climate Impact Assessment 2005*. Cambridge: Cambridge University Press.

Aleksey I (2008) Arktike opredelyat granitsy [Arctic borders will be determined]. *Rossiyskaya Gazeta*, 18 September.

AMAP (2010) *Assessment 2007: Oil and Gas in the Arctic, Effects and Potential Effects*. Vol 1. Oslo: Arctic Monitoring and Assessment Programme.

AMAP (2011) *Snow, Water, Ice and Permafrost in the Arctic (SWIPA): Climate Change and the Cryosphere*. Oslo: Arctic Monitoring and Assessment Programme. Available at: http://amap.no/swipa/CombinedReport.pdf (accessed 8 March 2019).

AMAP (2017a) *Adaptation Actions for a Changing Arctic: Perspectives from the Barents Area*. Oslo: Arctic Monitoring and Assessment Programme (AMAP).

AMAP (2017b) *Adaptation Actions for a Changing Arctic: Perspectives from the Bering-Chukchi-Beaufort Region*. Oslo: Arctic Monitoring and Assessment Programme (AMAP).

AMAP (2017c) *Snow, Water, Ice and Permafrost in the Arctic (SWIPA) 2017*. Oslo: Arctic Monitoring and Assessment Programme (AMAP).

Arctic Council (2013) *Arctic Council Task Force on Short-Lived Climate Forcers: Recommendations to Reduce Black Carbon and Methane Emissions to Slow Arctic Climate Change.* Available at: https://oaarchive.arctic-council.org/handle/11374/80 (accessed 8 March 2019).

Arctic Council (2016) *Arctic Resilience Report.* Carson M and Peterson G (eds). Stockholm: Stockholm Environment Institute and Stockholm Resilience Centre.

Baumeister C and Kilian L (2016) Lower oil prices and the US economy: Is this time different? *Brookings Papers on Economic Activity* 2016(2): 287–357. DOI: 10.1353/eca.2016.0029.

Bird KJ, Charpentier RR, Gautier DL et al. (2008) Circum-Arctic resource appraisal: Estimates of undiscovered oil and gas north of the Arctic circle. *US Geological Survey Fact Sheet 2008-3049.* US Geological Survey. Available at: http://pubs.usgs.gov/fs/2008/3049/fs2008-3049.pdf (accessed 10 August 2012).

Bjørst LR (2012) Climate testimonies and climate-crisis narratives: Inuit delegated to speak on behalf of the climate. *Acta Borealia* 29(1): 98–113. DOI: 10.1080/0800 3831.2012.678724.

Buurman T and Christensen M (2017) Governance and the changing Arctic: News framings in US newspapers from 2007 to 2015. In: *Conference on Communication and Environment (COCE)*, Leicester, 28 June–3 July.

Carrell S and van der Zee B (2010) Greenpeace 'shuts down' Arctic oil rig. *The Guardian*, 31 August. Available at: www.theguardian.com/environment/2010/aug/31/greenpeace-oil-rig-arctic (accessed 8 March 2019).

Cavazos-Guerra C, Lauer A and Rosenthal E (2017) Clean air and white ice: Governing black carbon emissions affecting the Arctic. In: Keil K and Knecht S (eds) *Governing Arctic Change: Global Perspectives.* London: Palgrave Macmillan UK, pp. 231–256. DOI: 10.1057/978-1-137-50884-3_12.

Cole D (2018) Climate plan or no, don't expect Alaska to stop drilling for new oil and gas. *Arctic Today*, 20 August. Available at: www.arctictoday.com/climate-plan-no-dont-expect-alaska-stop-drilling-new-oil-gas/ (accessed 23 August 2018).

Cole T and Cravez P (2004) *Blinded by Riches: The Permanent Funding Problem and the Prudhoe Bay Effect.* UA Research Summary no. 3. Anchorage, AL, US: Institute of Social and Economic Research, University of Alaska Anchorage.

Dale B and Kristoffersen B (2016) Imagining a postpetroleum Arctic. *Cultural Anthropology*, 29 July. Available at: https://culanth.org/fieldsights/943-imagining-a-postpetroleum-arctic (accessed 24 August 2018).

Dale B and Kristoffersen B (2018) Post-petroleum security in a changing Arctic: Narratives and trajectories towards viable futures. *Arctic Review on Law and Politics* 9: 244–261. DOI: 10.23865/arctic.v9.1251.

Dale B, Veland S and Hansen AM (2018) Petroleum as a challenge to arctic societies: Ontological security and the oil-driven 'push to the north'. *The Extractive Industries and Society* (online 23 October 2018); DOI: https://doi.org/10.1016/j.exis.2018.10.002.

Davenport C (2015) US will allow drilling for oil in Arctic Ocean. *New York Times*, 12 May.

Devyatkin DA, Suvorov RE and Sochenkov IV (2017) An information retrieval system for decision support: An Arctic-related mass media case study. *Scientific and Technical Information Processing* 44(5): 329–337. DOI: 10.3103/S0147688217050033.

Dokuchaev D (2012) Karachun na shelfe [Karachun on the shelf]. *Novaya Gazeta*, 5 September.

Dolgopolov N, Kondreva O and Fronon V (2011) Ukazaniye [Indication]. *Rossiyskaya Gazeta*, 15 September.

Funk M (2015) The wreck of the Kulluk. *New York Times*, 4 January. Available at: www.nytimes.com/2015/01/04/magazine/the-wreck-of-the-kulluk.html (accessed 14 September 2018).

Glomsrød S, Duhamine G and Aslaksen I (eds) (2015a) *The Economy of the North 2015*. Oslo: Statistics Norway. Available at: www.ssb.no/en/natur-og-miljo/artikler-og-publikasjoner/the-economy-of-the-north-2015 (accessed 23 August 2018).

Glomsrød S, Mäenpää I, Lindholt L et al. (2015b) Arctic economies within the Arctic nations. In: Glomsrød S, Duhamine G and Aslaksen I (eds) *The Economy of the North 2015*. Oslo, Norway: Statistics Norway, pp. 37–78. Available at: www.ssb.no/en/natur-og-miljo/artikler-og-publikasjoner/the-economy-of-the-north-2015 (accessed 23 August 2018).

Guettabi M (2016) What's happened to the Alaska economy since oil prices dropped? *Alaska Snapshot*: Report 1. Institute of Social and Economic Research, University of Alaska, Anchorage. Available at: https://scholarworks.alaska.edu:443/handle/11122/7804 (accessed 25 October 2018).

Guettabi M (2018) *How Do Oil Prices Influence Alaska and Other Energy-dependent States?* Anchorage, AL: University of Alaska Anchorage. Available at: www.alaskanomics.com/2018/10/how-do-oil-prices-influence-alaska-and-other-energy-dependent-states.html (accessed 8 March 2019)

Högselius P (2013) *Red Gas: Russia and the Origins of European Energy Dependence*. New York: Palgrave Macmillan.

IPCC (2018) Global Warming of 1.5°C. An IPCC Special Report on the impacts of global warming of 1.5°C above pre-industrial levels and related global greenhouse gas emission pathways, in the context of strengthening the global response to the threat of climate change, sustainable development, and efforts to eradicate poverty (Masson-Delmotte V, Zhai P, Pörtner HO, Roberts D, Skea J, Shukla PR, et al. (eds.)). Available at: www.ipcc.ch/report/sr15/ (accessed 23 October 2018).

Jensen LC and Hønneland G (2011) Framing the High North: Public discourses in Norway after 2000. *Acta Borealia* 28(1): 37–54. DOI: 10.1080/08003831.2011.575659.

Jensen Ø (2011) Current legal developments: The Barents Sea. *International Journal of Marine and Coastal Law* 26(1): 151–168. DOI: 10.1163/157180811X541422.

Johnson K (2015a) Oil company lease stirs revolt in green Seattle. *New York Times*. 14 March. Available at: www.nytimes.com/2015/03/14/us/oil-company-lease-stirs-revolt-in-green-seattle.html (accessed 14 September 2018).

Johnson K (2015b) Oil tumble grabs Alaska by the wallet. *New York Times*. 26 December.

Kartha S, Caney S, Dubash NK et al. (2018) Whose carbon is burnable? Equity considerations in the allocation of a 'right to extract'. *Climatic Change* 150(1–2): 117–129. DOI: 10.1007/s10584-018-2209-z.

Klimenko E, Nilsson AE and Christensen M (2019) *Narratives of conflict and cooperation in the Arctic and Russian media*. SIPRI Insights on Peace and Security. Stockholm: SIPRI.

Krauss C and Reed S (2015) Shell pulls plug on exploration in Alaska Arctic. *New York Times*, 29 September.

Larsen JN and Fondahl G (2014) *Arctic Human Development Report: Regional Processes and Global Challenges*. TemaNord 2014:567. Copenhagen: Nordic Council of Ministers.

Latynina Y (2008) Kosti Yukosa gniyut v tundre [Yukos' bones are rotting in the tundra]. *Novaya Gazeta*. 22 September.

Lazarus M and van Asselt H (2018) Fossil fuel supply and climate policy: Exploring the road less taken. *Climatic Change* 150(1–2): 1–13. DOI: 10.1007/s10584-018-2266-3.

Lindholt L and Glomsrød S (2015) Arctic petroleum extraction under climate policies. In: Glomsrød S, Duhaime G and Aslaksen, I (eds) *The Economy of the North 2015*. Oslo: Statistics Norway, pp. 79–84. Available at: www.ssb.no/en/natur-og-miljo/artikler-og-publikasjoner/the-economy-of-the-north-2015 (accessed 23 August 2018).

Loe J and Kelman I (2016) Arctic petroleum's community impacts: Local perceptions from Hammerfest, Norway. *Energy Research & Social Science* 16: 25–34. DOI: 10.1016/j.erss.2016.03.008.

MacFarquhar N (2015) Geysers, mushers and sled dogs vie with growth. *New York Times*, 5 April.

Monbiot G (2015) Keep fossil fuels in the ground to stop climate change. *The Guardian*, 10 March.

Mouawad J (2007) From the depth of the Arctic: Gas to heat homes in the US. *New York Times*, 9 October.

Nefidova N (2014) *Environmental Public Debate: In the Context of the Arctic in Russian and Norwegian Media*. Master's thesis. University of Oslo, Centre for Development and the Environment, Oslo, Norway. Available at: www.duo.uio.no/handle/10852/41797 (accessed 26 April 2018).

Nicol H (2013) Natural news, state discourses & the Canadian Arctic. In: Heininen L, Exner-Pirot H and Plouffe J (eds) *Arctic Yearbook 2013*. Northern Research Forum, pp. 211–236.

Nilsson AE (2018a) Creating a safe operating space for business: The changing role of Arctic governance. In: Wormbs N (ed.) *Competing Arctic Futures: Historical and Contemporary Perspectives*. Cham: Palgrave Macmillan, pp. 117–137.

Nilsson AE (2018b) The United States and the making of an Arctic nation. *Polar Record*: 1–13. DOI: 10.1017/S0032247418000219.

Nilsson AE and Filimonova N (2013) *Russian Interests in Oil and Gas Resources in the Barents Sea*. SEI Working Paper No. 2013-5. Stockholm: Stockholm Environment Institute. Available at: http://sei-international.org/publications?pid=2352 (accessed 1 October 2013).

Nilsson AE and Meek CL (2016) Learning to live with change. In: Carson M and Peterson G (eds) *Arctic Resilience Report*. Stockholm: Stockholm Environment Institute and Stockholm Resilience Centre, pp. 147–161.

Nilsson AE, Eklund N, Jürisoo M et al. (2019) Regional futures nested in global structures. In: Keskitalo EC (ed.) *The Politics of Arctic Resources: Change and Continuity in the 'Old North' of Northern Europe*. London: Routledge, pp. 221–262.

Norwegian Ministry of Foreign Affairs (2006) *Barents 2020: A Tool for Forward Looking High North Policy*. Oslo: Norwegian government. Available at: www.regjeringen.no/upload/UD/Vedlegg/barents2020e.pdf (accessed 6 August 2012).

Novaya Gazeta (2011) Shatkaya platforma (Shalky rig). 10 August.

Olsen E, Gjøsæter H, Røttingen I et al. (2007) The Norwegian ecosystem-based management plan for the Barents Sea. *ICES Journal of Marine Science: Journal du Conseil* 64(4): 599–602. DOI: 10.1093/icesjms/fsm005.

Pico S (2017) 48,300 Norwegian oil jobs have vanished. *ShippingWatch*. 20 January. Available at: https://shippingwatch.com/Offshore/article9305281.ece (accessed 25 October 2018).

Primakov Y (2015) Ne prosto rabotat', a znat' vo imja chego [Not just to work, but to know for what to work]. *Rossiyskaya Gazeta*, 15 January.

Prime Minister of Canada (2016) US-Canada Joint Statement on Climate, Energy, and Arctic Leadership. Available at: https://pm.gc.ca/eng/news/2016/03/10/us-canada-joint-statement-climate-energy-and-arctic-leadership (accessed 28 September 2018).

Reistad HH (2016) *Norway's Arctic Conundrum: Sustainable Development in the Norwegian Media Discourse.* Master's thesis in Sustainable Development 320. Department of Earth sciences, Uppsala University, Uppsala, Sweden. Available at: https://uu.diva-portal.org/smash/get/diva2:1039287/FULLTEXT01.pdf (accessed 7 March 2019).

Rosen Y (2018a) Alaska's climate action team suggests the state set up a carbon-pricing system. *Arctic Today*, 10 August. Available at: www.arctictoday.com/alaskas-climate-action-team-suggests-state-set-carbon-pricing-system/ (accessed 20 August 2018).

Rosen Y (2018b) At Inuit assembly, Alaska leader promotes oil development—on Inupiat terms. *Arctic Today*, 20 July. Available at: www.arctictoday.com/inuit-assembly-alaska-leader-promotes-oil-development-inupiat-terms/ (accessed 7 September 2018).

Rusbridger A (2015) Climate change: Why the Guardian is putting threat to Earth front and centre. *The Guardian*, 6 March. Available at: www.theguardian.com/environment/2015/mar/06/climate-change-guardian-threat-to-earth-alan-rusbridger (accessed 1 October 2018).

Sabitova N and Shavaleyeva C (2015) Oil and gas revenues of the Russian Federation: Trends and prospects. *Procedia Economics and Finance* 27. 22nd International Economic Conference of Sibiu 2015, IECS 2015 'Economic Prospects in the Context of Growing Global and Regional Interdependencies': 423–428. DOI: 10.1016/S2212-5671(15)01016-3.

Sellheim N (2013) The neglected tradition? The genesis of the EU seal products trade ban and commercial sealing. In: Alfredsson G and Koivurova T (eds) *The Yearbook of Polar Law*. Volume 5. Brill Nijhoff, pp. 417–450.

Sørensen CTN and Klimenko E (2017) *Emerging Chinese-Russian Cooperation in the Arctic: Possibilities and Constraints.* 46, SIPRI Policy Paper. Stockholm: SIPRI. Available at: www.sipri.org/sites/default/files/2017-06/emerging-chinese-russian-cooperation-arctic.pdf (accessed 8 March 2019).

Søviknes T (2018)Radio interview by Jens Möller. *Radiokorrespondenterna*. Swedish Broadcasting Cooperation. 2 September. Available at: https://sverigesradio.se/sida/avsnitt/1140957?programid=2946 (accessed 7 September 2018).

UNFCCC (2015) Paris Agreement. Adopted at the UNFCCC Conference of the Parties 21st session, Paris, 11 December 2015. FCCC/CP/2015/L.9. United Nations Framework Convention on Climate Change. Available at: https://unfccc.int/resource/docs/2015/cop21/eng/l09r01.pdf (accessed 12 February 2016).

United Nations (1992) United Nations Framework Convention on Climate Change (UNFCCC). Available at: http://unfccc.int/files/essential_background/background_publications_htmlpdf/application/pdf/conveng.pdf (accessed 6 September 2016).

Urry J (2013) *Societies beyond Oil: Oil Dregs and Social Futures.* London; New York: Zed Books.

US Energy Information Administration (2018a) US States: Rankings, Crude Oil Production, May 2018. Available at: www.eia.gov/state/rankings/#/series/46 (accessed 28 September 2018).

US Energy Information Administration (2018b) US States: Rankings, Total Energy Consumed per Capita, 2016. Available at: www.eia.gov/state/rankings/#/series/12 (accessed 28 September 2018).

Van Dusen J (2016) Nunavut, N.W.T. premiers slam Arctic drilling moratorium. CBC News. 22 December. Available at: www.cbc.ca/news/canada/north/nunavut-premier-slams-arctic-drilling-moratorium-1.3908037 (accessed 28 September 2018).

van Oort B, Bjørkan M and Klyuchnikova EM (2015) *Future Narratives for Two Locations in the Barents Region.* 2015:06, CICERO Report. CICERO Center for International Climate and Environmental Research: Oslo. Available at: http://hdl. handle.net/11250/2367371 (accessed 8 March 2019).

Warde P (2018) Constructing Arctic energy resources: The case of the Canadian North, 1921–1980. In: Wormbs N (ed.) *Competing Arctic Futures: Historical and Contemporary Perspectives.* Palgrave Macmillan, pp. 19–46.

Watt-Cloutier S, Fenge T and Crowley P (2006) Responding to global climate change: The view of the Inuit Circumpolar Conference on the Arctic Climate Impact Assessment. In: Rosentrater L (ed.) *2 Degrees Is Too Much! Evidence and Implications of Dangerous Climate Change in the Arctic.* Oslo: WWF, pp. 57–68. Available at: http://assets.panda. org/downloads/050129evidenceandimplicationshires.pdf (accessed 8 March 2019).

Zakharov M (2009) O polze prognozov [On the usefulness of forecasts]. *Novaya Gazeta.* 18 May.

6 Arctic geopolitics in times of transformation

On 8 October 2018, the Intergovernmental Panel on Climate Change (IPCC) issued a report on the consequences of a global temperature rise of 1.5 °C. The accompanying press release noted that: 'Limiting global warming to 1.5 °C would require rapid, far-reaching and unprecedented changes in all aspects of society' and that limiting warming 'could go hand in hand with ensuring a more sustainable and equitable society' (IPCC, 2018a). In the Arctic, warming happens two to three times faster than in the rest of the world, and the IPCC report states:

> There is *high confidence* that the probability of a sea-ice-free Arctic Ocean during summer is substantially lower at global warming of 1.5°C when compared to 2°C. With 1.5°C of global warming, one sea-ice-free Arctic summer is projected per century. This likelihood is increased to at least one per decade with 2°C global warming.
>
> (IPCC, 2018b: 10)

Less than two weeks after the IPCC report was published, 2,000 people gathered for the Annual Arctic Circle Assembly in Reykjavik to continue discussions on the 'challenges and opportunities' presented by the new Arctic that is emerging. The participants came from all over the world, including Asia and the Pacific Islands. Iceland's former president and chair of the Arctic Circle, Ólafur Ragnar Grímsson, noted that 'Our neighborhood has become a global playing field' (Grímsson, 2018). Depledge and Dodds (2017) argue that this annual event has become a 'bazaar' at which actors that have been marginalized by the formal circumpolar governance of the Arctic Council have a free hand to sell their ideas and imaginaries of the Arctic and its future. While the impacts of climate change served as an unavoidable context at this meeting, as in previous meetings of the Arctic Circle, the focus of the discussion remained more on the Arctic as a global space on to which visions of new infrastructure and business opportunities can be projected. The 2018 Arctic Circle Assembly included several sessions about China's Belt and Road Initiative, as well as a lavish reception funded by the Chinese Embassy featuring Chinese stage performers. Japan and the EU also attended the assembly for the first time. The Arctic Council had an unassuming stand among other institutional exhibitors. Its voice in plenary sessions was limited to a short presentation of the priorities of the upcoming Icelandic Chair, which will commence in the spring of 2019.

While, throughout the 2018 Arctic Circle Assembly, the various media were mainly represented by specialist journalists who focus on the Arctic, Arctic change has occupied an increasingly large mainstream media space over the past decade, as discussed in previous chapters. Through news channels and popular venues, such as online and mobile platforms, audio-visual exhibitions, literature and film, narratives of environmental transformations, geopolitical tensions and future scenarios have been digitally mediated and remediated. In this process, conceptions of 'actual' space in the geographic sense and 'virtual' space in the mediated sense in the role of geopolitics have shaped each other. According to Christensen et al. (2018: 1):

> The actual and virtual sites and locales (e.g. museums, electronic media space, film, music, archives, etc.) where narrative interventions materialize constitute *spaces of narrativity. Narrated space* (such as 'the ocean' in a broad, and 'the Arctic' or 'the Ozone layer' in specific senses) signifies the site of environmental transformation.

Events such as the Arctic Circle Assembly demonstrate how realities and urgencies present in *narrated space* and discourses put forth in *spaces of narrativity* can merge and collide. As the geographic alterations in the Arctic (i.e. narrated space) due to a warming climate affect both the near and the far, the contours of narrated space expand to include the presence of new entrants such as Japan, China and South Korea. In parallel with this process, the range of spaces of narrativity on to or through which the local and global, corporate and state, and public and private imaginaries are projected drastically increase. In each case, the space in question both informs and is informed by geopolitics and power interplays.

Media ecologies and Arctic spaces of narrativity

Spaces of narrativity are subject to certain dynamics, and these are rapidly changing. While mainstream media and certain forms and genres of visuality and textuality more easily find central spaces of presence and a wide circulation, others such as small-budget or experimental productions and outlets find a voice from the margins. As is noted in Chapter 2, in a changing media ecology, integrative perspectives are needed to grasp the dynamic interplays between actual and virtual spaces and scales. To return to our discussion in the introductory chapter, two notions are especially relevant in the context of the Arctic: disintermediation and media ecology. While it has long been speculated that elite gatekeeping will disappear and content created horizontally will disintermediate and dwindle the role of the mainstream media as the primary political intermediaries, other voices have argued that traditional media will retain their significance and influence as they feed from and into these alternative outlets (Aday et al., 2013). A perspective based on media ecology that takes account of the complexities of such interplays in mediatized environments avoids such binaries as traditional/big versus alternative/small media. The media ecology perspective instead conceptualizes the media as 'an environment that surrounds the subjects and models their cognitive and perceptual system', and the disintermediation perspective 'looks at the interactions between media, as if they were species of an ecosystem' (Scolari, 2012: 209–210). Such perspectives go beyond the reductionist understanding of the media only as digital forums and make it possible to understand offline forms of communication, such as art, music, performativity, activism and orality, within that ecology.

In the case of the Arctic, as revealed throughout this book, big-media interest is still mostly linked to event-based reporting, and their in-depth, extensive stories are often the result of expensive 'parachute journalism'. While such journalism is costly and difficult to offer on a regular basis, the resulting news stories and reports can reach a wide audience, especially when remediated via social and small media. There has also been a growing diversity in the topics and themes of mainstream media reporting about the Arctic, as well as an increasing use by international news media of specialized online news sites such as *Arctic Today*, which provide 'on-site' and locally contextualized reporting.

Conflicting, cooperating and invisible narratives

Over the past decade, stories about climate change have been interwoven with other recurring and sometimes conflicting narratives of the north, such as the race for resources, most notably access to offshore oil and gas, the Arctic as an arena for potential geopolitical conflict, and the Arctic as a unique environment that must be protected from the impacts of industrial development. Meanwhile, Arctic states have tried hard to defuse conflicts on respect of both environmental goals and competition for resources. The emphasis has been on international *cooperation* in the region, with the Arctic cast as a zone of peace in a world of increasing conflict. The Arctic should be safe.

The emphasis on cooperation and managing potential conflicts within existing governance mechanisms has been further underscored by the signing of several legally binding agreements, such as when the five Arctic Ocean littoral states— Canada, Norway, Russia, Denmark and the United States—and the five global fishing giants in the northern hemisphere—Iceland, Japan, South Korea, China and the European Union—agreed a moratorium on fishing in the Central Arctic Ocean in early October 2018 (e.g. Harvey, 2018; Sevunts, 2018). This moratorium did not stem from an Arctic Council initiative. It therefore signals something different from earlier agreements on search and rescue, oil spill prevention and scientific cooperation. Even if these earlier cooperation agreements were not Arctic Council agreements per se, since this regional body does not have the power to enter into legally binding agreements, they nonetheless placed Arctic regional cooperation centre stage. By contrast, the Arctic Ocean fishing moratorium illustrates how Arctic governance has become increasingly framed as a global concern. It follows a similar logic to the Ilulissat Declaration of 2008 (Arctic Ocean Conference, 2008), where the Arctic littoral states emphasized the UN Convention on the Law of the Sea (UNCLOS) as the major governance mechanism for resolving interstate disputes in the Arctic—a premise that has since been widely accepted not just in the Arctic but beyond the region.

Another attempt to manage potential conflicts in the Arctic is the emphasis in Arctic Council initiatives on preventing pollution and ensuring best practices in industrial development. This was evident in the Arctic Council's Kiruna Vision of 2013, which highlighted the notion of 'a safe Arctic' (Arctic Council, 2013; Nilsson, 2018). In such discussions, states are often framed as the most legitimate actors. This is apparent from analyses of discourses about sustainability in the Arctic, where Gad and Strandsbjerg (2019) highlight 'how nation states and their concerns with economic development take on a privileged position' and that the concept of sustainability is used 'to reconfirm states and markets as central identities for the future development of the Arctic'. In the process, goal conflicts become 'a technical problematique' of making the exploitation of natural resources as benign as possible. Meanwhile, the conflict between the exploitation of Arctic oil and gas and the need to halt global warming by cutting emissions of greenhouse gases is neglected, as discussed in Chapter 5. Moreover, the narrative leaves no space to talk about the geopolitical implications of transforming the Arctic region in a post-petroleum future. In media reporting, geopolitical issues are mainly linked to access to petroleum resources and there is an implicit assumption that current consumption levels must continue, in stark contrast to the IPCC's conclusions.

The attempts to reframe potential conflicts between industrial development and environmental goals into cooperative ventures on sustainability have not been quite as successful as the handling of potential conflicts over rights to Arctic resources. One major reason for this is that environmental organizations continually work to boost a counter narrative: that industrial development in the Arctic, especially the exploitation of oil and gas, could spell disaster for the Arctic environment. Their spectacularly staged events, such as the kayak protest in Seattle harbour calling attention to Shell's activities in the Chukchi Sea and the attempt by the *Arctic Sunrise* to scale Gazprom's rig in the Kara Sea, are designed to get media attention. In the mainstream media, however, these stories were covered as 'events' and attention usually dissipated once the event itself was over. Mainstream media's coverage of larger environmental events of global scale displays the same pattern. For instance, while there were representatives of vulnerable communities, trade unions and Arctic indigenous groups present in the front lines of the People's Climate March in 2014, news reporting focused on the size and spectacle and celebrity speeches. Similarly, the voices and images of the Saami and other Arctic communities did not feature in international news during the negotiation of the Paris Agreement in December 2015 (Christensen et al., 2014; Russell, 2016).

In the shadow of larger conflicts with global dimensions, and despite the growing multiplicity of Arctic topics, the stories of social transformations in the Arctic in recent years remain underreported by the mainstream media.

For the four million people who live in the region, the most important impacts of both climate change and growing industrial development are those that are felt at the local level and affect daily livelihoods, as well as local and subnational regional economies. Some stories on local impacts fit well within a global framing, not least the impacts of warming on indigenous cultures. In such cases, indigenous voices are often subsumed into narratives on global environmental change or victimhood. Other potential news stories do not fit as neatly into global narratives. These include the local implications of industrial development, whether positive or negative, as seen from a local perspective. These are also the stories that do not immediately fit into the oft-repeated mantra: 'What happens in the Arctic doesn't stay in the Arctic'. Journalists and editors of major mainstream media outlets might not see them as newsworthy, given that their readers' interest in the Arctic is steered by the global impacts of changes in the northern polar region. Thus, while climatic and economic impacts are felt and narrated globally, what happens in the locales of the Arctic *does* stay there unless avenues are available for voicing local concerns to broad audiences.

Regional mediation networks

Non-simplistic narration of local life and realities often has difficulties competing with existing frames and norms for space and visibility in a mediatized social landscape. Nonetheless, local stories still have a natural place in local media reporting. Until recently, such local media reports were divided by language, national context and distribution channels, but the growth of a web-supported circumpolar media landscape is changing this picture. As discussed in Chapter 2, the sharing of stories across news outlets in different parts of the region is creating a setting where journalists and editors specializing in Arctic issues connect these stories to each other and into a larger picture. For example, at a panel on Arctic journalism at the 2018 Arctic Circle, Eilis Quinn from *Eye on the Arctic* told the audience about how she had placed similar stories about the impacts of mining and extraction on reindeer herding and on caribou together on the website, and that she worked with the aim of making similar links with other issues. In the process, the phenomena become something more than disconnected local conflicts. Instead, regional patterns become visible, creating the potential for the sharing of experiences across different localities and countries. This speaks to the potential power of alliances between small-scale media outlets to create new networks of '*comingtogetherness*' for people in the Arctic.

It is possible to compare the new connections that are starting to emerge from circumpolar media reporting with a process that began in the scientific assessments of environmental change in the Arctic carried out by the working groups of the Arctic Council and its predecessor the Arctic Environmental Protection Strategy. Such assessments put insights from national observations and studies into a circumpolar context and created a new understanding of the Arctic as a region. One specific example is how the Arctic Climate Impact Assessment mapped local observations of the impacts of climate change by indigenous people onto a circumpolar picture in order to create a new and convincing story about climate change being real in the here and now (Huntington and Fox, 2005: 68–69). Over the years, such coordination has created a picture of a region where the impacts of human activities on the environment cannot be ignored. These include the impacts not only of climate change, but also of long-range pollution, and they have framed the Arctic as part of a global environmental change context rather than an exceptional and pristine region.

In recent decades, circumpolar scientific networks have grown to now include a range of disciplines as well as major efforts to produce interdisciplinary work. This growth has been supported partly by the Arctic Council and its assessments, but also by scientific organizations such as the International Arctic Science Committee (IASC) and the International Arctic Social Science Association (IASSA), as well as by initiatives such as the International Polar Year 2007–2008 and the recurrent strategic planning of Arctic research by the International Conference on Arctic Research Planning (ICARP). In 2017, the legal agreement on scientific cooperation in the Arctic provided further institutional support for a circumpolar perspective on Arctic research.

The regional and local media networks that are now emerging involve fewer people and lack formal institutional support. Their activities have, at times, also been circumscribed by economic challenges or by the political will to suppress certain discourses. They are, nonetheless, part and parcel of the circumpolar civil society community that has emerged during the recent decades of circumpolar state cooperation in the Arctic Council. Together with similar networks based on myriad scientific collaborations along with the formal and informal networks that now connect indigenous peoples across the Arctic, they represent a circumpolar society that is independent of formal political cooperation. These networks have also created the basis for a regional Arctic communicative realm that might not be immediately recognizable as geopolitical because they have lesser presence and influence in formal international politics, but nonetheless have geopolitical implications. Their potential power lies more in the extent to which they will be able to influence overarching mediations of the region and its futures than in formal power in international regimes.

Transformation challenges

In the coming years, people living in the Arctic will face many challenges that will require open and candid discussion about both immediate and long-term consequences. Some of these issues are linked to political decisions at the local, national and circumpolar levels but many will be influenced by global developments, including climate change and shifts in the markets for raw material and energy resources. As the IPCC has emphasized, the need to limit global warming requires 'rapid, far-reaching and unprecedented changes in all aspects of society' (IPCC, 2018a). The implications for the Arctic go beyond shifting to greener technologies to meet local energy needs, which may be comparatively easy, given the new technologies that are now emerging and the local benefits of cheaper and more reliable energy sources. Much more challenging will be how the global shifts that are necessary will affect the demand for Arctic resources.

For communities and countries that are currently dependent on income from oil and gas, the necessary shift will require diversification of the economy. In areas with the potential for renewable energy in the form of wind power or hydro-power, conflicts over land use are already an issue. At the Arctic Circle Assembly in 2018, indigenous peoples voiced concerns that, once again, they would be forced to carry a disproportionate burden of a global challenge (Schreiber, 2018). Communities and countries where mining is economically important might also face the consequences of shifts in global demand, including a potential push for new mines to access minerals that are important to the green technologies. In the wake of such increasing demand, there is likely to be increased competition for land. Furthermore, even if efforts to radically halt global emissions of greenhouse gases are successful, Arctic warming will continue for some time. There is thus also a need for adaptation to new climate conditions and to share knowledge and experience of how to best navigate the new situation.

Media with a circumpolar reach could potentially play an important role in supporting networks and knowledge-sharing across the Arctic. However, there are major risks that media will not be able to fulfil this potential. These include political tensions that prevent various media from working freely as well as the economic vulnerability that affects what reporters and media organiza-tions can do, even when ambitions are high. Furthermore, the fragmentation of the media landscape, narrowcasting and the growth of social media that cater only to specific groups complicates the picture, as does mistrust of established media channels. Fragmentation and lack of trust have geopolitical implications if diverging discourses are pitched against each other to fuel or escalate con-flicts rather than help resolve disagreements. Furthermore, the lack of nuance and space for criticism increases the risk that recurring global and national nar-ratives will maintain their hegemony and leave little space for the discussion of alternative futures based on local aspirations and priorities.

Complex geopolitics

In formal regional cooperation, the Arctic Council continues to function despite political differences between the Arctic states. Adherence to applying UNCLOS norms to resolve potential conflicts in the Arctic has the support of Arctic and non-Arctic states alike, despite the reluctance of the United States to ratify it. Nonetheless, while discourses and displays of cooperation and peace continue to define the region, growing military presence raises concerns about darker sides of geopolitics (Wezeman, 2016). These include Russian military activities in the Arctic, combined with Russia's strong economic and geostrategic interests, which have both regional and global implications. In addition, while NATO does not have a unified strategy for the Arctic, a NATO military exercise in Norway in October and November 2018—Trident Junction—*de facto* involved the Arctic. Trident Junction is the largest NATO exercise in decades and the first one north of the Arctic Circle in three decades. While NATO maintains that this activity is not meant as a military threat to Russia, it was partly a response to Russia's increased submarine presence in the region. The Trident Junction exercise has been heavily mediated in international news, which warrants a reflection on the geo and politics in geopolitics.

Understandings of the role of geopolitics in international dynamics have oscillated between classical and critical schools of the discipline, and Arctic change in combination with how the region is narrated provides fertile ground for comparing and contrasting the different schools (Eklund and van der Watt, 2017; Wegge and Keil, 2018). Furthermore, these trends highlight a need to understand the role of media and mediation, including the dialectical relationship between the discursive realm and materiality. As discussed in Chapter 2, a conception of mediation that construes the media beyond discourse and representation and places centre stage the materialities of media presence and connectivity would alleviate the rift between the theoretical dichotomies of hard and soft power. In this context, geoeconomic relations and how they intertwine with the increased globalization in finance, economy, political and cultural domains become important. In the Arctic, the supremacy of an economic focus is most visible. While climate change and its regional and global consequences occupy a large space in both political and media discourses, forums such as the Arctic Circle, Arctic Frontiers and the Arctic Economic Council are home to convergent and contested business ideas and ideals involving the Arctic and its resources. In such discourses, climate change is discussed in relation to opportunities more than threats. An important step in harnessing these opportunities is support of new infrastructure, including communication infrastructures that will inevitably influence the Arctic at various levels, including the mediation of various discourses and their reach.

What define the geopolitics of the Arctic today are the dynamic interactions of social, environmental and technological systems and how these affect discourse formation and narratives. While military and political presence and displays of power have implications that extend *beyond* the region to potentially include non-Arctic powers and their interests, discourses that emphasize justice and environmental violence, as highlighted by post-structuralist conceptions of geopolitics (critical geopolitics, popular geopolitics, feminist geopolitics), *traverse* the region across scales through mediated circuits. The increased fibre optic connectivity of the region as well as the emergence of regional and local media outlets and alliances thus complement and have the potential to challenge global, often reductionist, mediations of the Arctic. While local and regional voices and their international political impact might not today be at the level where they can powerfully impact the future of the Arctic, increased connectivity brings with it some promise and hope of more nuance.

Concluding remarks

There are limits to the potential for regional international governance in cases where strong state interests are at stake. When challenged by uncertainty because of the impacts of climate change and the growing interest from actors outside the region, the framing of the region has shifted from a circumpolar focus to a combination of reasserted state interests and new assertions of the Arctic as a global space. In this process, local voices have had difficulties projecting their concerns. However, the period of peaceful cooperation in the Arctic over the past three decades has created civil society and media networks across national borders in the region, including connections among indigenous peoples and among researchers. Collaboration across local media outlets and their growing significance in larger media ecologies can give local stories circumpolar and global reach, and increase the potential for more nuanced narration of the Arctic. In a geoeconomic and geopolitical context of contested spaces and futures, the 'regional' no longer resides in specific governance mechanisms but in social networks enabled by increased communications capacities and power to influence.

As discussed throughout this volume, multiple factors shape both the material and discursive dimensions of regions, providing an exceptional opportunity to rethink the role of geography in politics. According to classical geopolitics, material conditions and physical geography reign supreme in influencing state power and status in international relations and politics. By contrast, critical accounts of geopolitics challenge the determinism inherent in a focus on physical geography and place centre stage the role of discursive and cultural factors. One lesson from the analysis in this book is that discourses, physical geographies and other materialities, such as infrastructure, interact in a dynamic manner. A second lesson is linked to the continuing expansion of infrastructure and communication networks in the region. These economically and politically driven investments have supported the emergence of regional, small-scale media capacity to communicate local voices globally, voices and visions that were previously missing from the geopolitical scene. Added to this are other forms of discursive interventions such as films and art that convey messages and visions about the localities in the Arctic, thanks to reduced production and circulation costs in the current media ecology. While the visible and envisaged impacts of climate change remain a major driver of Arctic geopolitics, with new actors, new interests and new concerns, in addition to Arctic states that have increased stakes both commercially and in relation to military and security issues, the role of communication infrastructure and networks should not be underestimated in influencing future visions and priorities. Geography and space matter in actual and virtual terms as the Arctic continues to transform.

References

Aday S, Farrell H, Freelon D et al. (2013) Watching from afar: Media consumption patterns around the Arab Spring. *American Behavioral Scientist* 57(7): 899–919. DOI: 10.1177/0002764213479373.

Arctic Council (2013) Vision for the Arctic. Kiruna, Sweden, 15 May 2013. Available at: www.arctic-council.org/index.php/en/document-archive/category/425-main-documents-from-kiruna-ministerial-meeting (accessed 17 February 2014).

Arctic Ocean Conference (2008) The Ilulissat Declaration. Arctic Ocean Conference, Ilulissat Greenland, 27–28 May 2008.

Christensen M, Robin L and Möller N (2014) Climate change show and tell. *Le Monde Diplomatique* (English Edition), November 2014.

Christensen M, Åberg A, Lidström S et al. (2018) Environmental themes in popular narratives. *Environmental Communication* 12(1): 1–6. DOI: 10.1080/17524032.2018.1421802.

Depledge D and Dodds K (2017) Bazaar governance: Situating the Arctic Circle. In: Keil K and Knecht S (eds) *Governing Arctic Change*. London: Palgrave Macmillan UK, pp. 141–160. DOI: 10.1057/978-1-137-50884-3.

Eklund N and van der Watt L-M (2017) Refracting (geo)political choices in the Arctic. *The Polar Journal* 7(1): 86–103. DOI: 10.1080/2154896X.2017.1337334.

Gad UP and Strandsbjerg J (eds) (2019) Conclusion: Sustainability reconfiguring identity, space, and time. In: *The Politics of Sustainability in the Arctic: Reconfiguring Identity, Space, and Time*. Routledge studies in sustainability. Abingdon, Oxon; New York: Routledge, pp. 242–254.

Grímsson, ÓR (2018) Speech at the concluding plenary of the 2018 Arctic Circle Assembly, Reykjavik, Iceland. Personal notes by Nilsson, AE.

Harvey F (2018) Commercial fishing banned across much of the Arctic. *The Guardian*, 3 October. Available at: www.theguardian.com/environment/2018/oct/03/commercial-fishing-banned-across-much-of-the-arctic (accessed 23 October 2018).

Huntington H and Fox S (2005) The changing Arctic: Indigenous perspectives. In: *Arctic Climate Impact Assessment (ACIA)*. Cambridge, UK: Cambridge University Press, pp. 61–98.

IPCC (2018a) Press release: Summary for Policymakers of IPCC Special Report on Global Warming of 1.5°C approved by governments. Available at: http://ipcc.ch/news_and_events/pr_181008_P48_spm.shtml (accessed 23 October 2018).

IPCC (2018b) Global Warming of 1.5°C. An IPCC Special Report on the impacts of global warming of 1.5°C above pre-industrial levels and related global greenhouse gas emission pathways, in the context of strengthening the global response to the threat of climate change, sustainable development, and efforts to eradicate poverty (Masson-Delmotte V, Zhai P, Pörtner HO, Roberts D, Skea J, Shukla PR et al. (eds)). Available at: www.ipcc.ch/report/sr15/ (accessed 23 October 2018).

Nilsson AE (2018) Creating a safe operating space for business: The changing role of Arctic governance. In: Wormbs N (ed.) *Competing Arctic Futures: Historical and Contemporary Perspectives*. Cham: Palgrave Macmillan, pp. 117–137.

Russell A (2016) *Journalism as Activism: Recoding Media Power*. John Wiley & Sons.

Schreiber M (2018) The Arctic's indigenous peoples bear a disproportionate burden of the world's response to climate change, leaders say. *Arctic Today*, 9 November. Available at: www.arctictoday.com/arctics-indigenous-peoples-bear-disproportionate-burden-worlds-response-climate-change-leaders-say/ (accessed 20 November 2018).

Scolari CA (2012) Media ecology: Exploring the metaphor to expand the theory. *Communication Theory* 22(2): 204–225. DOI: 10.1111/j.1468-2885.2012.01404.x.

Sevunts L (2018) Binding agreement on Arctic fisheries moratorium officially signed by EU and nine countries. *Eye on the Arctic*, 3 October. Available at: www.rcinet.ca/eye-on-the-arctic/2018/10/03/fishing-fisheries-moratorium-agreement-arctic-ocean-europe-union-canada-greenland-bouffard/ (accessed 23 October 2018).

Wegge N and Keil K (2018) Between classical and critical geopolitics in a changing Arctic. *Polar Geography* 41(2): 87–106. DOI: 10.1080/1088937X.2018.1455755.

Wezeman S (2016) *Military Capabilities in the Arctic: A New Cold War in the High North?* SIPRI Background paper. Stockholm: SIPRI.

Index

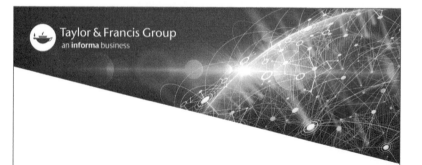